Food Authentication

Food Authentication

Techniques, Trends and Emerging Approaches

Special Issue Editor

Raúl González-Domínguez

MDPI • Basel • Beijing • Wuhan • Barcelona • Belgrade • Manchester • Tokyo • Cluj • Tianjin

Special Issue Editor
Raúl González-Domínguez
University of Huelva
Spain

Editorial Office
MDPI
St. Alban-Anlage 66
4052 Basel, Switzerland

This is a reprint of articles from the Special Issue published online in the open access journal *Foods* (ISSN 2304-8158) (available at: https://www.mdpi.com/journal/foods/special_issues/Food_authentication_Techniques_Trends_Emerging_Approaches).

For citation purposes, cite each article independently as indicated on the article page online and as indicated below:

LastName, A.A.; LastName, B.B.; LastName, C.C. Article Title. *Journal Name* **Year**, *Article Number*, Page Range.

ISBN 978-3-03928-748-2 (Pbk)
ISBN 978-3-03928-749-9 (PDF)

Contents

About the Special Issue Editor . . . . **vii**

Preface to "Food Authentication" . . . . **ix**

Raúl González-Domínguez
Food Authentication: Techniques, Trends and Emerging Approaches
Reprinted from: *Foods* **2020**, *9*, 346, doi:10.3390/foods9030346 . . . . **1**

Mizuki Morisasa, Tomohiko Sato, Keisuke Kimura, Tsukasa Mori and Naoko Goto-Inoue
Application of Matrix-Assisted Laser Desorption/Ionization Mass Spectrometry Imaging for
Food Analysis
Reprinted from: *Foods* **2019**, *8*, 633, doi:10.3390/foods8120633 . . . . **5**

**Vassilios K. Karabagias, Ioannis K. Karabagias, Artemis Louppis, Anastasia Badeka, Michael
G. Kontominas and Chara Papastephanou**
Valorization of Prickly Pear Juice Geographical Origin Based on Mineral and Volatile
Compound Contents Using LDA
Reprinted from: *Foods* **2019**, *8*, 123, doi:10.3390/foods8040123 . . . . **23**

Marilena E. Dasenaki, Sofia K. Drakopoulou, Reza Aalizadeh and Nikolaos S. Thomaidis
Targeted and Untargeted Metabolomics as an Enhanced Tool for the Detection of Pomegranate
Juice Adulteration
Reprinted from: *Foods* **2019**, *8*, 212, doi:10.3390/foods8060212 . . . . **39**

Fien Minnens, Niels Lucas Luijckx and Wim Verbeke
Food Supply Chain Stakeholders' Perspectives on Sharing Information to Detect and Prevent
Food Integrity Issues
Reprinted from: *Foods* **2019**, *8*, 225, doi:10.3390/foods8060225 . . . . **59**

**Ra úl González-Domínguez, Ana Sayago, María Teresa Morales
and Ángeles Fernández-Recamales**
Assessment of Virgin Olive Oil Adulteration by a Rapid Luminescent Method
Reprinted from: *Foods* **2019**, *8*, 287, doi:10.3390/foods8080287 . . . . **75**

Guillem Campmajó, Laura Cayero, Javier Saurina and Oscar Núñez
Classification of Hen Eggs by HPLC-UV Fingerprinting and Chemometric Methods
Reprinted from: *Foods* **2019**, *8*, 310, doi:10.3390/foods8080310 . . . . **85**

**Enrique Durán-Guerrero, Mónica Schwarz, M. Ángeles Fernández-Recamales,
Carmelo G. Barroso and Remedios Castro**
Characterization and Differentiation of Spanish Vinegars from Jerez and Condado de Huelva
Protected Designations of Origin
Reprinted from: *Foods* **2019**, *8*, 341, doi:10.3390/foods8080341 . . . . **95**

**Spyridon Papapetros, Artemis Louppis, Ioanna Kosma, Stavros Kontakos, Anastasia Badeka,
Chara Papastephanou and Michael G. Kontominas**
Physicochemical, Spectroscopic and Chromatographic Analyses in Combination with
Chemometrics for the Discrimination of Four Sweet Cherry Cultivars Grown in
Northern Greece
Reprinted from: *Foods* **2019**, *8*, 442, doi:10.3390/foods8100442 . . . . **107**

**Luciana Piarulli, Michele Antonio Savoia, Francesca Taranto, Nunzio D'Agostino,
Ruggiero Sardaro, Stefania Girone, Susanna Gadaleta, Vincenzo Fucili,
Claudio De Giovanni, Cinzia Montemurro, Antonella Pasqualone and Valentina Fanelli**
A Robust DNA Isolation Protocol from Filtered Commercial Olive Oil for
PCR-Based Fingerprinting
Reprinted from: *Foods* **2019**, *8*, 462, doi:10.3390/foods8100462 . . . . **123**

**Sanae Bikrani, Ana M. Jiménez-Carvelo, Mounir Nechar, M. Gracia Bagur-González,
Badredine Souhail and Luis Cuadros-Rodríguez**
Authentication of the Geographical Origin of Margarines and Fat-Spread Products from Liquid
Chromatographic UV-Absorption Fingerprints and Chemometrics
Reprinted from: *Foods* **2019**, *8*, 588, doi:10.3390/foods8110588 . . . . **137**

Ángela Alcazar Rueda, José Marcos Jurado, Fernando de Pablos and Manuel León-Camacho
Differentiation between Ripening Stages of Iberian Dry-Cured Ham According to the Free
Amino Acids Content
Reprinted from: *Foods* **2020**, *9*, 82, doi:10.3390/foods9010082 . . . . **149**

Raúl González-Domínguez, Ana Sayago, Ikram Akhatou and Ángeles Fernández-Recamales
Multi-Chemical Profiling of Strawberry as a Traceability Tool to Investigate the Effect of Cultivar
and Cultivation Conditions
Reprinted from: *Foods* **2020**, *9*, 96, doi:10.3390/foods9010096 . . . . **163**

About the Special Issue Editor

Raúl González-Domínguez received his PhD in Chemistry in 2015 (University of Huelva, Spain) and then moved to the University of Barcelona as a postdoctoral researcher. His research interests are mainly focused on the development of metabolomics tools based on mass spectrometry and chromatographic-based approaches for metabolite profiling, as well as their application in biomedicine (e.g., age-related diseases, metabolic disorders), nutrition (e.g., discovery of food intake biomarkers, impact of diet on health), and food research (e.g., food authentication and traceability). To date, he is the author of 60 research and review articles in peer-reviewed international journals, 5 book chapters, and one patent. He has participated in 12 international projects and 14 Spanish national projects funded through competitive calls, as well as in 5 R&D contracts with public and private entities. Dr. González-Domínguez has also been involved in the direction and supervision of more than 35 research projects for under- and postgraduate students.

 *foods*

Editorial

Food Authentication: Techniques, Trends and Emerging Approaches

Raúl González-Domínguez [1,2]

[1] Department of Chemistry, Faculty of Experimental Sciences, University of Huelva, 21007 Huelva, Spain;
    raul.gonzalez@dqcm.uhu.es; Tel.: +34-959-219-975
[2] International Campus of Excellence CeiA3, University of Huelva, 21007 Huelva, Spain

Received: 12 March 2020; Accepted: 13 March 2020; Published: 17 March 2020

Multiple factors can directly influence the chemical composition of foods and, consequently, their organoleptic, nutritional and bioactive properties, including the geographical origin, the variety or breed, as well as the conditions of cultivation, breeding and/or feeding, among others. Therefore, there is a great interest in the development of accurate, robust and high-throughput analytical methods to guarantee the authenticity and traceability of foods. For these purposes, a large number of sensorial, physical and chemical approaches can be used, which must normally be combined with advanced statistical tools. In this vein, the aim of the Special Issue "Food Authentication: Techniques, Trends and Emerging Approaches" was to gather original research papers and review articles focused on the development and application of analytical techniques and emerging approaches in food authentication. This Special Issue is comprised of 12 valuable scientific contributions, including one review article and 11 original research works, dealing with the authentication of foods with great commercial value, such as olive oil, Iberian ham or fruits, among others.

Morisasa et al. reviewed the potential of matrix-assisted laser desorption/ionization mass spectrometry (MALDI-MS) imaging as a valuable technique to determine small metabolites in food tissue sections without requiring purification, extraction, separation or labeling processes [1]. They highlight that MALDI-MS can be employed not only to identify the nutritional content of foods, but also to investigate their geographical origin for improved traceability, food safety and breed enhancement, among other applications. However, the authors also emphasize that further technical improvements are needed, especially to overcome sensitivity issues.

Several research articles reported the application of chromatographic-based analytical approaches for profiling different analytes as possible chemical descriptors for authenticity and traceability purposes. Rueda et al. determined the free amino acid content of Iberian dry-cured hams to differentiate among three ripening stages: postsalting, drying and cellar [2]. For this purpose, they employed gas chromatography coupled to mass spectrometry (GC-MS) and flame ionization detector (GC-FID) to identify and quantify 18 amino acids. Alanine, tyrosine, glutamine, proline, 2-aminobutanoic acid, cysteine and valine were found to be the most differentiating amino acids between the ripening stages by using principal component analysis (PCA) and linear discriminant analysis (LDA), which could be therefore used to predict the curing time. Volatile profiling by GC-MS (alcohols, aldehydes, hydrocarbons, terpenoids) combined with mineral content determination (25 macro- and microminerals) was employed to classify prickly pear juice samples from the Peloponnese Peninsula according to the geographical origin [3]. Multivariate analysis demonstrated that seven minerals and 21 volatile compounds provide satisfactory classification rates, and furthermore, mineral content of soil samples was satisfactorily correlated with mineral levels detected in fruit juices. Similarly, Papapetros et al. also investigated volatile and mineral profiles, combined with other conventional physicochemical and spectroscopic determinations (e.g., acidity, total phenolic content, sugars), to differentiate sweet cherry samples grown in northern Greece according to the botanical origin [4]. Results evidenced that individual datasets provide acceptable but not satisfactory classification rates, whereas their combination leads to improved classification models. In another study, two Spanish Protected Designation of Origin vinegar samples were analyzed for polyphenol and volatile content by liquid and gas chromatography approaches, respectively [5]. Multivariate data analysis demonstrated clear differences between vinegars with regard to their polyphenolic content, and to a lesser extent, in the volatile fraction. Authors proposed that these differences should be mainly due to varietal and geographical factors, since vinegar manufacturing and ageing processes are similar in both regions. To achieve a comprehensive characterization of the chemical composition of strawberry fruits, González-Domínguez et al. applied a multitargeted profiling approach to determine multiple compounds related to sensory and health characteristics of this berry fruit, including sugars, organic acids, polyphenols and mineral elements [6]. Then, several complementary pattern recognition procedures were employed to discriminate strawberry varieties grown under different climatic and agronomic conditions. Anthocyanins, phenolic acids, sucrose and malic acid showed significant differences among cultivars, while climatic conditions and the cultivation system were responsible for changes in polyphenol contents. In this vein, metabolomics has also been proposed as a powerful screening tool for authenticity assessment [7]. Targeted and nontargeted metabolomics approaches were used to detect pomegranate juice adulteration with apple and red grape juice. This methodology allowed distinguishing adulteration to levels below 1%, and 80 potential biomarkers were identified (e.g., anthocyanins, flavonoids).

The use of spectroscopic methods for food authenticity research has also been reported in some research articles published in this Special Issue, as detailed below. Campmajó et al. described the application of high-performance liquid chromatography with ultraviolet detection (HPLC-UV) to detect "fingerprints" for the classification of hen eggs according to their production method: organic, free-range, barn or caged [8]. Multivariate modeling enabled satisfactory discrimination rates, especially for the distinction among organic and nonorganic eggs. However, perfect classification of the four egg groups was not achieved, so authors proposed that future research lines could include the evaluation of egg yolk instead of the whole egg, and the use of fluorescence detection as a more selective technique. Using a similar analytical approach based on LC-UV fingerprinting, Bikrani et al. were able to differentiate margarines and fat-spread-related products from different geographical origins from Spain and Morocco [9]. Several multivariate chemometrics tools were compared, with partial least squares-discriminant analysis (PLS-DA) being the statistical strategy that provided the best performance. In this line, luminescence also demonstrated a great potential to characterize edible oils and detect adulterations in a rapid way [10]. In this work, a regression model based on five luminescent frequencies, associated with minor oil components, was designed and validated for detecting virgin olive oil adulteration with hazelnut oil.

Piarulli et al. developed a robust DNA-isolation protocol from extra virgin olive oil (EVOO) for subsequent polymerase chain reaction (PCR)-based fingerprinting [11]. This method was then successfully applied for genetic tagging of filtered EVOOs of unknown origin. Finally, the work by Minnens et al. aimed to investigate attitudes towards a food integrity information sharing system (FI-ISS) among stakeholders in the European food supply chain [12].

In summary, the Special Issue "Food Authentication: Techniques, Trends and Emerging Approaches" evidences the great importance of developing novel analytical approaches to define accurate and reproducible indicators for food authenticity and traceability. At the same time, as suggested by several authors, the application of advanced chemometrics approaches is also essential to achieve robust results, with the aim of characterizing food composition, discovering potential markers (e.g., adulteration) and obtaining satisfactory classification models.

Conflicts of Interest: The author declare no conflict of interest.

References

1. Morisasa, M.; Sato, T.; Kimura, K.; Mori, T.; Goto-Inoue, N. Application of Matrix-Assisted Laser Desorption/Ionization Mass Spectrometry Imaging for Food Analysis. *Foods* **2019**, *8*, 633. [CrossRef] [PubMed]
2. Rueda, Á.A.; Jurado, J.M.; de Pablos, F.; León-Camacho, M. Differentiation between Ripening Stages of Iberian Dry-Cured Ham According to the Free Amino Acids Content. *Foods* **2020**, *9*, 82. [CrossRef] [PubMed]
3. Karabagias, V.K.; Karabagias, I.K.; Louppis, A.; Badeka, A.; Kontominas, M.G.; Papastephanou, C. Valorization of Prickly Pear Juice Geographical Origin Based on Mineral and Volatile Compound Contents Using LDA. *Foods* **2019**, *8*, 123. [CrossRef] [PubMed]
4. Papapetros, S.; Louppis, A.; Kosma, I.; Kontakos, S.; Badeka, A.; Papastephanou, C.; Kontominas, M.G. Physicochemical, Spectroscopic and Chromatographic Analyses in Combination with Chemometrics for the Discrimination of Four Sweet Cherry Cultivars Grown in Northern Greece. *Foods* **2019**, *8*, 442. [CrossRef] [PubMed]
5. Durán-Guerrero, E.; Schwarz, M.; Fernández-Recamales, M.Á.; Barroso, C.G.; Castro, R. Characterization and Differentiation of Spanish Vinegars from Jerez and Condado de Huelva Protected Designations of Origin. *Foods* **2019**, *8*, 341. [CrossRef] [PubMed]
6. González-Domínguez, R.; Sayago, A.; Akhatou, I.; Fernández-Recamales, Á. Multi-Chemical Profiling of Strawberry as a Traceability Tool to Investigate the Effect of Cultivar and Cultivation Conditions. *Foods* **2020**, *9*, 96. [CrossRef] [PubMed]
7. Dasenaki, M.E.; Drakopoulou, S.K.; Aalizadeh, R.; Thomaidis, N.S. Targeted and Untargeted Metabolomics as an Enhanced Tool for the Detection of Pomegranate Juice Adulteration. *Foods* **2019**, *8*, 212. [CrossRef] [PubMed]
8. Campmajó, G.; Cayero, L.; Saurina, J.; Núñez, O. Classification of Hen Eggs by HPLC-UV Fingerprinting and Chemometric Methods. *Foods* **2019**, *8*, 310. [CrossRef] [PubMed]
9. Bikrani, S.; Jiménez-Carvelo, A.M.; Nechar, M.; Bagur-González, M.G.; Souhail, B.; Cuadros-Rodríguez, L. Authentication of the Geographical Origin of Margarines and Fat-Spread Products from Liquid Chromatographic UV-Absorption Fingerprints and Chemometrics. *Foods* **2019**, *8*, 588. [CrossRef] [PubMed]
10. González-Domínguez, R.; Sayago, A.; Morales, M.T.; Fernández-Recamales, Á. Assessment of Virgin Olive Oil Adulteration by a Rapid Luminescent Method. *Foods* **2019**, *8*, 287. [CrossRef] [PubMed]
11. Piarulli, L.; Savoia, M.A.; Taranto, F.; D'Agostino, N.; Sardaro, R.; Girone, S.; Gadaleta, S.; Fucili, V.; De Giovanni, C.; Montemurro, C.; et al. A Robust DNA Isolation Protocol from Filtered Commercial Olive Oil for PCR-Based Fingerprinting. *Foods* **2019**, *8*, 462. [CrossRef] [PubMed]
12. Minnens, F.; Lucas Luijckx, N.; Verbeke, W. Food Supply Chain Stakeholders' Perspectives on Sharing Information to Detect and Prevent Food Integrity Issues. *Foods* **2019**, *8*, 225. [CrossRef] [PubMed]

Review

Application of Matrix-Assisted Laser Desorption/Ionization Mass Spectrometry Imaging for Food Analysis

Mizuki Morisasa, Tomohiko Sato, Keisuke Kimura, Tsukasa Mori and Naoko Goto-Inoue *

Department of Marine Science and Resources, College of Bioresource Sciences, Nihon University, 1866 Kameino, Fujisawa, Kanagawa 252-0880, Japan; xx1mizu9xx@gmail.com (M.M.); sato310.volley@gmail.com (T.S.); mucunguiyou412@gmail.com (K.K.); mori.tsukasa@nihon-u.ac.jp (T.M.)
* Correspondence: inoue.naoko@nihon-u.ac.jp; Tel.: +81-46-684-3681

Received: 25 October 2019; Accepted: 28 November 2019; Published: 2 December 2019

Abstract: Food contains various compounds, and there are many methods available to analyze each of these components. However, the large amounts of low-molecular-weight metabolites in food, such as amino acids, organic acids, vitamins, lipids, and toxins, make it difficult to analyze the spatial distribution of these molecules. Matrix-assisted laser desorption/ionization mass spectrometry (MALDI-MS) imaging is a two-dimensional ionization technology that allows the detection of small metabolites in tissue sections without requiring purification, extraction, separation, or labeling. The application of MALDI-MS imaging in food analysis improves the visualization of these compounds to identify not only the nutritional content but also the geographical origin of the food. In this review, we provide an overview of some recent applications of MALDI-MS imaging, demonstrating the advantages and prospects of this technology compared to conventional approaches. Further development and enhancement of MALDI-MS imaging is expected to offer great benefits to consumers, researchers, and food producers with respect to breeding improvement, traceability, the development of value-added foods, and improved safety assessments.

Keywords: MALDI-MS imaging; amino acids; lipids; neuropeptides; nutrition factor

---

1. Introduction

Food ingredients contain a wide variety of nutritional components such as carbohydrates, proteins, peptides, lipids, minerals, vitamins, amino acids, and organic acids. In addition to the intentionally included ingredients, food can also contain contaminants such as pesticide residue, the residue of pharmaceuticals given to animals, mycotoxins, food additives, and carcinogenic substances introduced during food processing. Therefore, from the point of view of food safety, it is very important to verify all constituents in food. The diverse physical properties of these constituents require varied methods of analysis for detection, and these ingredients generally must first be purified for qualitative and quantitative analyses. Gas chromatography-mass spectrometry (GC-MS) and high-performance liquid chromatography-mass spectrometry (HPLC-MS) are commonly used to analyze food components. GC-MS is used for the detection and identification of volatile organic compounds such as amino acids, polyols, and vitamins, which are commonly derivatized [1]. HPLC-MS, however, is used to analyze higher molecular weight polar compounds [2,3]. Therefore, the application scope of these methods is wide and versatile. However, GC-MS and HPLC-MS are not well-suited for the analysis of the spatial distribution of compounds in food. In addition to the amounts of these compounds, determining their localization could provide useful information for food safety and for plant breeding and food processing applications. Conventional imaging techniques such as infrared spectrometry (IR) [4] and magnetic resonance imaging (MRI) [5] are also widely applied to food analysis. However, compounds from the spectrum are difficult to identify with IR, and MRI is not effective to identify substances with a low water content.

MS-based imaging utilizing the principle of matrix-assisted laser desorption/ionization (MALDI) is a relatively new imaging method for small metabolites [6–8], lipids [9–12], and proteins (peptides) [13–15]. MALDI-MS imaging is a two-dimensional analysis method that can detect intact molecules within tissue sections without requiring extraction, purification, separation, or labeling, and is the most applicable method owing to its ability to detect a wide range of molecules. The spatial resolution of MALDI-MS, depending on the laser radiation interval, can be set over a wide range of 5–200 µm, which is sufficient to obtain the molecular distribution from a single cell. Because MALDI-MS imaging can detect all molecules that undergo ionization, it has attracted significant attention as a non-target analysis method. Unlike conventional imaging methods such as immunohistochemistry, MALDI-MS imaging does not require the labeling of target molecules before analyses, making it one of the most powerful and convenient tools to screen molecules that show characteristic localization, especially in the medical field. The tandem mass spectrometry feature used directly on tissue sections allows for structure identification at a region of interest. In this review, we focus on recent applications of MALDI-MS imaging for food analyses, highlighting the great potential of this technique to improve quality control and food safety.

2. History of MALDI-MS Imaging Applications

The concept of MALDI-MS imaging was first introduced in the early 2000s. Caprioli et al. [16] performed direct ionization of proteins using rat pancreatic tissue sections, and Stoeckli et al. [17] demonstrated protein localization in the mouse cerebrum. At that time, the spatial resolution was still at a visual level (mm), but technological innovation improved this to 10 µm, and high-speed analysis with laser frequencies at 200–1000 Hz was realized in the 2010s [6]. These advances allowed for the capture of single cell-specific molecular distributions with MALDI-MS imaging.

MALDI-MS imaging was initially developed to analyze protein localization. At this time, there was a great demand for the detection of post-translational modifications according to a mass difference or the localization of hormone-like substances for which specific antibodies are difficult to obtain. However, this was not an easy task owing to quantitative limitations of these molecules. In contrast, lipids and very small molecules such as organic acids and nucleic acids were confirmed to be favorable molecules for detection with MALDI-MS imaging [18–20]. MALDI-MS imaging emerged in the medial field, especially in the field of cancer science to gain a better understanding of the metabolite dynamics during tumorigenesis [21], and has also been applied to the food field [22]. In food science, MALDI-MS imaging is now widely used to determine the localization of specific substances of interest such as sugars, amino acids, lipids, and polyphenols [23–25], or to screen non-target molecular dynamics in breeding or dosing applications. Thus, MALDI-MS imaging is used not only to clarify the localization of substances in food but also to ensure food safety.

MALDI-MS imaging has emerged as a valuable tool for the visualization of low-molecular-weight metabolites, but sometimes the matrix itself contributes to ionization interference. For this reason, other ionization methods such as desorption electrospray ionization mass spectrometry (DESI-MS) [26,27] and secondary ion mass spectrometry (SIMS) [28] are also used for MS imaging. DESI-MS [29] was introduced in 2004 as an ambient ionization method to directly ionize solid-phase samples at atmospheric pressure. One of the main advantages of DESI-MS is that since it does not require matrix spraying, it induces minimal damage to the sample without complicated matrix interference [30], thereby enabling the detection of minor components. Moreover, DESI-MS does not require the use of a special coating with indium thin oxide (ITO) to the slide glass, and the samples can be simultaneously applied to both imaging and histochemical analyses, allowing for accurate correlation analysis between molecular signatures and the histological state. In some cases, DESI-MS can ionize nonpolar compounds such as carotenes, which are abundantly present in plants and difficult to ionize with MALDI-MS. These properties have made DESI-MS a valuable tool for within-tissue detection of the spatial distribution of specialized metabolites [31], the visualization of plant metabolites, and investigations of their biological roles [32]. In contrast, the main advantage of SIMS is the ability to measure the spatial localization of molecules with high spatial resolution. MALDI imaging has a resolution of several tens of micrometers, whereas that of SIMS can be 200 nm or less. Although the samples used for SIMS also do not require any special surface treatment, the samples might be lost since SIMS can be a destructive analysis. Nevertheless, the improved resolution has now made it possible to image biomolecules at the sub-cellular level with this technique [28]. Those three ionization methods, MALDI, DESI, and SIMS, are the most used analytical imaging techniques in food sciences.

The number of published articles with the key term "mass spectrometry imaging" in the PubMed database has gradually increased in the last 20 years. Only 90 papers on the topic were published up to 1995, whereas more than 3000 papers applying this technology have been published from 2016 to 2019. With respect to its application in food sciences, there were only 10 papers retrieved from a search with the key term "mass spectrometry imaging food" up to 2000. However, the number of papers doubled every 5 years thereafter, increasing to 15 in 2001–2005 and to 36 in 2006–2010; moreover, this number has been increasing rapidly in the last decade, with 240 papers reported since 2016 in this field. Many of these studies were aimed at using MS imaging to ensure the reliability of food and to increase the value of the food (Figure 1).

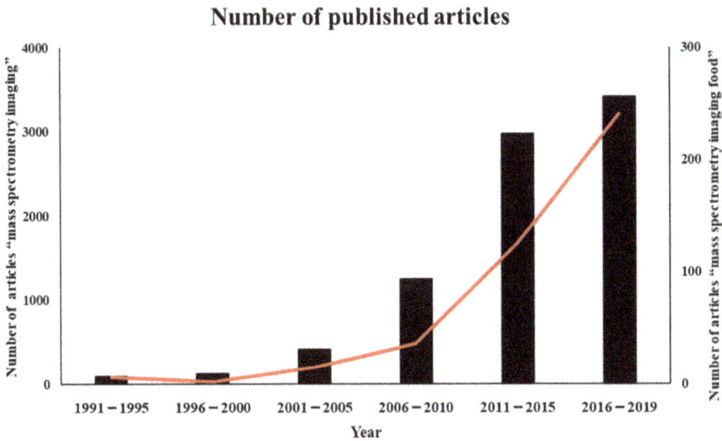

Figure 1. Publication trend (number of published articles) related to mass spectrometry imaging (black bars) and mass spectrometry imaging of food (red line).

3. Sample Pretreatment for MALDI MS-Imaging

A schematic of the protocol for MALDI-MS imaging is shown in Figure 2. Here, we focus on the most important experimental steps such as sectioning, pretreatment of the section, and choice of matrix (e.g., 2,5-dihydroxybenzoic acid, α-cyano-4-hydroxycinnamic acid, 9-aminoacridine) and method for matrix application (e.g., spraying, deposition, and sublimation). To obtain useful results with MALDI-MS imaging, the sample pretreatment step is arguably the most important overall. In particular, the sample type, size, thickness, matrix, method of coating matrix, and other related factors need to be predetermined. Table 1 summarizes the recent applications of MALDI-MS imaging for food samples, demonstrating a wide variety of target molecules associated with equally diverse preparation methods. The optimal conditions for sample preparation need to be determined according to the sample. Here, we focus on the key aspects and related methods to prepare a sample to ensure reliable MALDI-MS imaging data.

Table 1. Overview of the application of matrix-assisted laser desorption/ionization mass spectrometry (MALDI-MS) imaging for food science and related fields.

Sample	Target Molecules	Sample Preparation Sample Type Thickness Embedding	Matrices	Reference
Soya leaf, stem	Mesotorione, azoxystrobin	Freeze-drying - -	CHCA	[8]
Strawberry fruit skin	Sucrose, fructose, glucose, citric acid	Fresh 0.2–0.5 mm with a sharp utility knife -	DHB	[33]
Wheat grain	Glucose-6-phosphate, sucrose	Frozen - Ice	CHCA	[7]
Wheat stem	Oligosaccharides	Freeze-drying 50 μm -	CHCA	[34]
Ginger rhizome (*Zingiber officinale*)	6-gingerol, monoterpene	Fresh 0.2 mm -	-	[6]
Eggplant	GABA, nicotinic acid, arginine, 2-aminobenzoic acid, citric acid, saccharides	Frozen 14 μm -	DHB	[23]
Blue swimming crab (*Portunus pelagicus*)	Phospholipids, triacylglycerols	Frozen 14 μm 2% CMC	DHB	[35]
Rice seed	Phospholipids, α-tocopherol, arginine, γ-oryzanol, phytic acid	Frozen 8 μm with adhesive film (Kawamoto method) 2% CMC	DHB	[36]
Beef meat	Lipids	Frozen 8 μm -	DHB	[37]
Penaeus monodon	Neuropeptides	Frozen 5 μm Paraffin	CHCA	[13]

Table 1. *Cont.*

Sample	Target Molecules	Sample Preparation Sample Type Thickness Embedding	Matrices	Reference
Capsicum annuum	Capsaicin	Frozen 70 μm -	CHCA	[25]
Black rice seed	Lysophosphatidylcholine, phosphatidylcholine, anthocyanins	Frozen 10 μm with adhesive film (Kawamoto method) 2% CMC	DHB	[38]
Camelina sativa seed transgenic	Lipids	Frozen 30–50 μm 10% gelatin	DHB	[10]
Potato (*Solanum tuberosum* L.)	α-solanine, α-chaconine	Frozen - -	CHCA	[39]
Wheat (*Triticum aestivum* L.)	Polysaccharides	Frozen 60 μm -	DHB	[40]
Tomato fruit (*S. lycopersicum* L.)	Organic acid, amino acid nucleotides, caffeic acid	Frozen 10 μm OCT compound	DHB, 9-AA	[41]
Rice (*Oryza sativa* L.)	Cytokinin, abscisic acid	Frozen 50 μm Ice	CHCA	[42]
Cucumber	Triterpenes	Frozen 50 μm -	-	[43]
Maize seed (*Zea mays*)	Triacylglycerols, amino acids	Frozen 10 μm -	DAN, DHB, 9-AA	[44]
Oilseed rape (*Brassica napus*)	Lipids	Frozen 30 μm -	DHB	[45]
Strawberry	Anthocyanins, sugars, organic acids	Frozen 80 μm 2% CMC	DHB	[24]

Table 1. *Cont.*

Sample	Target Molecules	Sample Preparation Sample Type Thickness Embedding	Matrices	Reference
Red sea bream (*Pagrus major*)	Lipids	Frozen 15 μm -	DHB	[46]
Grain (*Triticum aestivum* L.)	Hemicelluloses	Frozen 80 μm -	DMA, DHB	[47]
Ham	Peptide	Frozen 12 μm -	CHCA	[48]
Apple	Soluble carbohydrate	Fresh 20 μm -	CHCA, DHB	[49]
Nightshades	Alkaloids	Frozen 40 μm Ice	DHB	[50]
Pork chop	Lipids	Frozen 10 μm -	CHCA, DHB	[9]

GABA, gamma-aminobutyric acid; CMC, carboxymethyl cellulose; OCT, optimal cutting temperature; CHCA, α-cyano-4-hydroxycinnamic acid; DAN, 1,5-diaminonaphthalene; DHB, 2,5-dihydroxybenzonic acid; DMA, *N*,*N*-dimethylaniline; 9-AA, 9-aminoacridine.

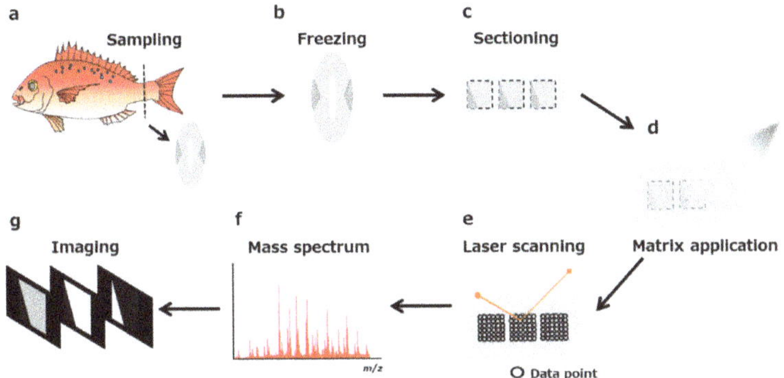

Figure 2. Scheme of matrix-assisted laser desorption/ionization mass spectrometry matrix-assisted laser desorption/ionization mass spectrometry (MALDI-MS) imaging. (**a**) Tissue sampling. (**b**) Preparation of a fresh-frozen sample. (**c**) Sectioning. (**d**) Application of the matrix. (**e**) Laser scanning. (**f**) Procurement of the mass spectrum. (**g**) Visualization of the ion distribution of molecules.

3.1. Sample Storage

The biological tissue samples used for MALDI-MS imaging require storage at −80 °C to maintain the intact form and spatial organization of the biomolecules in the samples. The most favorable tissues for this purpose are fresh-frozen tissues, which can be prepared with various methods such as using powered dry ice, liquid nitrogen, or liquid nitrogen-chilled isopentane, among others [51]. In addition to frozen tissues, formalin-fixed, paraffin-embedded (FFPE) tissues sections could be applied to MALDI-MS imaging. However, FFPE tissues sections are completely stripped off lipophilic molecules after the deparaffinization step [52]; therefore, these sections have been adapted to detect proteins or peptides.

3.2. Embedding

Small-sized tissues and high-water content samples are hard to cut into appropriate sections without mounting. However, the use of typical embedding agents such as an optimal cutting temperature (OCT) compound must be avoided for samples destined for MALDI-MS imaging since the molecules derived from these agents can introduce ionization interference with respect to the biomolecules of interest [53]. Therefore, carboxymethyl cellulose (CMC) is recommended as the embedding material of tissue samples for MALDI-MS imaging, in addition to the use of 2% sodium CMC as an alternative embedding compound [54]. Khatib-Shahidi et al. [55] detected the drug and metabolite distributions in whole-body tissue sections at various time points following drug administration using CMC-embedded tissues. This was the first report indicating that CMC-embedded tissues can be used for MALDI-MS imaging. Especially, foods and/or plant with high water content sometimes need embedding to maintain their shape [24,36]. This approach is also applicable for the discovery of the localization of nutritional factors in plants.

3.3. Sectioning

The ionization efficiency is partly dependent on the thickness of the tissue section. In general, 5–20-μm-thick sections are prepared for the analysis of low-molecular-weight molecules. The use of thinner tissue sections (2–5 μm in thickness) are recommended for the analysis of high-molecular-weight molecules (3–21 kDa). We recommend the use of an ITO-coated glass slide for thaw-mounting of the sections because these transparent slides enable microscopic observation of the section after MALDI-MS imaging.

One of the major challenges of MALDI-MS imaging is maintaining the original shape of the tissue during the preparation of sections, which is particularly difficult when a section is created from fragile, and/or hard samples. High-quality sections can be made by creating a section using an adhesive film [56], which is suitable for attaching a tissue as a section, and is then attached to MALDI target plates or ITO-coated glass slides. However, this is associated with the severe drawback of reduced ionization efficiency compared to that obtained when using frozen sections on an ITO-coated glass slide [57]. There is also a method to transfer only the tissue from the film to the glass slide using an ultraviolet light source, but this requires a very advanced technique and specialized equipment [58]. Recently, Nakabayashi et al. [59] reported a technique to directly transfer a prepared section from a film to ITO-coated glass slides, which improves sample preparation for plant tissues with a high water content.

3.4. Sample Pretreatment (Washing, Digestion)

For protein imaging, it is essential to first wash out the lipophilic molecules on the tissue sections. The typical washing procedure is performed with a gradient of ethanol reagents. Finally, the sections are dried in a vacuum desiccator for 15 min [60]. For the detection of digested peptides, treatment with protease is also performed [61]. As mentioned, FFPE sections are used for protein and/or peptide detection [40]. In addition to FFPE sections, frozen sections could by applied for peptide imaging. Heijs et al. [62] conducted a comprehensive study of the mouse brain proteome using MALDI-MS imaging. With the treatment of complementary proteases, they were able to identify 5337 peptides from the MALDI-MS imaging results, corresponding to 1198 proteins. In addition to the common proteases such as trypsin, pepsin, and elastase [63], an enzyme for cutting N-linked glycans (peptide N-glycosidase F; PNGase F) has also been used to analyze post-translational modifications [64]. N-glycans play a fundamental role in many molecular and cellular processes and are established biomarkers of diseases [65,66]. MALDI-MS imaging with PNGase F—based tissue digestion has been successfully applied for the analysis of N-glycans in tissue microarrays [67]. For quantitative analysis, stable isotope-labeled internal standards are sprayed onto the sections [68].

3.5. Derivatization for Minor Targets

Ionized molecules are limited to abundant molecules and/or molecules with high ionization efficiency. Therefore, it is more challenging to detect minor components with conventional sample preparation. MALDI-MS imaging can be applied to overcome this problem through the development of a derivatization approach. Derivatization has long been used for quantitative analysis by GC-MS and HPLC-MS as it can enhance the ionization efficiency of molecules that are difficult to ionize. Due to the improved efficiency of derivatization, it has become a more common application in the field of MALDI. A representative example of such minor components is neurotransmitters. Derivatization converts low-molecular-weight endogenous primary amines into readily-ionized species of relatively high molecular weight, which enables the simultaneous imaging of tyrosine, tryptamine, tyramine, phenethylamine, dopamine, 3-methoxytryramine, serotonin, gamma-aminobutyric acid (GABA), and glutamate, along with other endogenous primary amines, in a single MALDI-MS image. These technologies have now made it possible to measure the distributions of a wide range of primary amines with high resolution [69]. We must also take into account the possibility of the lateral drift of molecules due to the reagents. To prevent the dispersion of molecules with these solvent, a previous study used an automatic sprayer to fix spray conditions.

3.6. Matrix Selection

The selection of an appropriate matrix is undoubtedly the most important step in optimizing the MALDI-MS conditions for a given application. In particular, the purity, concentration, and solubility are important considerations to adapt to the chemical properties of the target compounds. In addition to the type of matrix, there are various methods to apply a matrix to the sections. The commonly used matrices for MALDI-MS imaging are listed in Table 2. The matrix plays a central role in MALDI-MS soft ionization [70,71], in which the biomolecules are softly ionized in the co-crystal with the matrix, which absorbs the laser beam energy to protect the biomolecules from the disruptive energy. Protonated ions ($[M + H]^+$) or deprotonated ions ($[M - H]^-$) are generally detected, and sodium adduct ions ($[M + Na]^+$) and potassium adduct ions ($[M + K]^+$) are often observed in the analyses of biological samples. Thus, the choice of an appropriate matrix is essential to obtain meaningful biomolecule images [72,73]. The choice of a matrix used for MALDI-MS imaging depends on the chemical properties and mass range of the analytes of interest. In general, 2,5-dihydroxybenzoic acid and 9-aminoacridine are considered suitable materials for lipids and small metabolites.

Table 2. Common matrix types used in matrix-assisted laser desorption/ionization mass spectrometry (MALDI-MS) imaging according to the sample type.

Matrices	Sample
9-Aminoacridine (9-AA)	lipids, metabolites
Sinapinic acid (SA)	peptides, proteins
Nicotinic acid (NA)	nucleotide
2,5-Dihydroxybenzoin acid (DHB)	lipids, glycopeptide, polymer
3-Amino-4-hydroxybezoic acid (AHBA)	glycan
α-Cyano-4-hydroxycinnamic acid (CHCA)	peptides, proteins
1,5-Diaminonapthalene (DAN)	lipids
t3-Indolacrylic acid (IAA)	polymer, aromatic
2-(4-Hydroxyphenylazo)-benzoic acid (HABA)	polymer
3-Aminoquinoline (3AQ)	glycan
Picolinic acid (PA)	nucleotide
Anthranilic acid (ANA)	nucleotide
3-Hydroxypicolinic acid (3HPA)	nucleotide
5-Chlorosalycilic acid (5CSA)	polymer
Dihydroxyacetone phosphate (DHAP)	lipids, glycan

3.7. Matrix Coating

In MALDI-MS imaging, the analytes must be co-crystallized with the matrices to be appropriately ionized. Therefore, we must minimize the crystal size of matrices that influence the spatial resolution of molecules. There are various methods available to coat the matrix onto the section, including deposition, spraying, and sublimation. The method of matrix application also influences the analyte extraction efficiency. Spraying is the most frequently used method in MALDI-MS imaging [74], and can coat the entire tissue section with relatively small crystals that are homogeneously dispersed within a short time without any special equipment. However, this method requires skillful operation to manipulate the numerous parameters with a manually operated airbrush.

To overcome the problem of reproducibility associated with spraying conditions, an automatic matrix sprayer system can be used, which was previously reported to result in approximately double the number of metabolites detected compared to that detected when using the manual airbrush method [75]. Thus, the use of an automated system could enhance the sensitivity of molecule detection. An explanation for this effect is that the TM-Sprayer, an automatic matrix sprayer system, sprays the solution at a high temperature, thereby forming more homogeneous matrix crystals on the tissue than possible with an airbrush [75,76].

4. Localization of Small Metabolites in Foods

MALDI-MS imaging is now becoming widely applied for localization analysis of molecules in food, which is one of the simplest methods of MALDI-MS imaging, revealing characteristic substances with various approaches. Some of these studies have focused on the localization of nutritional factors. GABA plays various beneficial roles such as controlling plant growth and resistance to oxidative stress, and is one of the most widely recognized ingredients of functional foods. Although the precise localization of GABA had not been clarified because of the lack of a specific probe, MALDI-MS imaging, for the first time, enabled visualization of its characteristic localization in the seeds of eggplants.

The determination of the fine localization of GABA in crops would be useful for plant breeding and food processing [23]. In the same manner, small molecules in plant tissues were previously analyzed to identify these localizations [34,41]

Fruit taste is largely dependent on the balance and content of sweetness and sourness. Sweetness and acidity are determined by sugar (especially glucose, fructose, and sucrose) and organic acid (citric acid and malic acid), respectively; however, their spatial localization remains undetermined. The application of MALDI-MS to strawberries revealed different substances in each part of the fruit. For example, glucose and fructose were distributed throughout the strawberry section, whereas sucrose was predominantly distributed on the top side of the cortical tissue and in the vascular bundles. This study further showed that the sweetness of each part of the strawberry differs according to the constituent substance of that part. Thus, MALDI-MS imaging not only clarified the characteristic patterns of metabolism in strawberries but also provided reference information to guide future strawberry breeding improvements [24].

Solanine and chaconine, contained in potatoes, are steroidal alkaloid glycosides that are known to be toxic to the human body; consuming large amounts of these toxins results in typical symptoms of nausea and abdominal pain. The total amount of toxins contained in potato buds has been determined by separation analysis methods such as liquid chromatography, but there are few methods available to visualize their distribution in foodstuffs. Tissue distribution determined by MALDI-MS imaging revealed the presence of solanine and chaconine on the bud and perimeter of persimmons. This approach therefore provides visually accessible data for experts and for consumers, who can intuitively understand the location of targeted components [39].

Capsaicin is a pungent ingredient of capsicum fruits, which can be roughly divided into the pericarp, placenta, and seed regions. MALDI-MS imaging demonstrated that capsaicin was localized to the surface of the placenta and pericarp but was scarcely present in the seed. From these results, it was considered that capsaicin is likely present in the pericarp to protect the fruit from external enemies, and that a large amount of capsaicin is produced on the surface of the placenta to protect the seed [25].

5. Breed Improvement with MALDI-MS Imaging-Based Localization Analysis

In addition to issues of food safety and quality control, identifying the distribution and contents of various food components can also provide valuable information for plant breeding and aquaculture. Rice is a crop that has undergone many breed improvements to increase its nutritional quality. Zaima et al. [36] used MALDI-MS imaging to demonstrate how the localization of substances (metabolites and minerals) in rice can be used to investigate changes occurring with breed improvement. In this study, the researchers detected lysophosphatidylcholine within the endosperm, α-tocopherol within the scutellum, and phosphatidycholine, γ-oryzanol, and phytic acid in the bran. In particular, this was the first study to analyze the characteristic distributions of α-tocopherol, γ-oryzanol, and phytic acid, as three important nutritional components. In addition, the localization of arginine was different as a result of breeding, with various distribution patterns among rice types and growth stages.

Anthocyanin is present in various forms with respect to the glycosides. The localization of seven types of anthocyanin monoglycosides and two types of anthocyanin diglycosides was determined in black rice and blueberries, demonstrating that the distribution varied among the different structures of the foods [38]. Furthermore, using MALDI-MS imaging, which does not require extraction and concentration, the researchers could detect fragile molecules including two types of anthocyanin pentosides, which are difficult to detect by conventional HPLC owing to their unstable structure.

MALDI-MS imaging has also been applied in the field of aquaculture. Sroyraya et al. [35] visualized the metabolites of the eyestalk of the blue swimming club, and Piyachat et al. [13] showed the distribution of neuropeptides in shrimp. We also analyzed the metabolic molecules of wild and farmed red sea bream to identify species-specific metabolic differences. Because the complete genome sequence of red sea bream is not yet complete, it has been very difficult to find molecular markers of wild and farmed fish for their respective identification. However, using MALDI-MS imaging, we were able to discover some lipid molecules that differed between wild and farmed fish. Furthermore, the specific distribution of anserine, a well-known nutrition factor, in farmed red sea bream could be detected (Figure 3) [46].

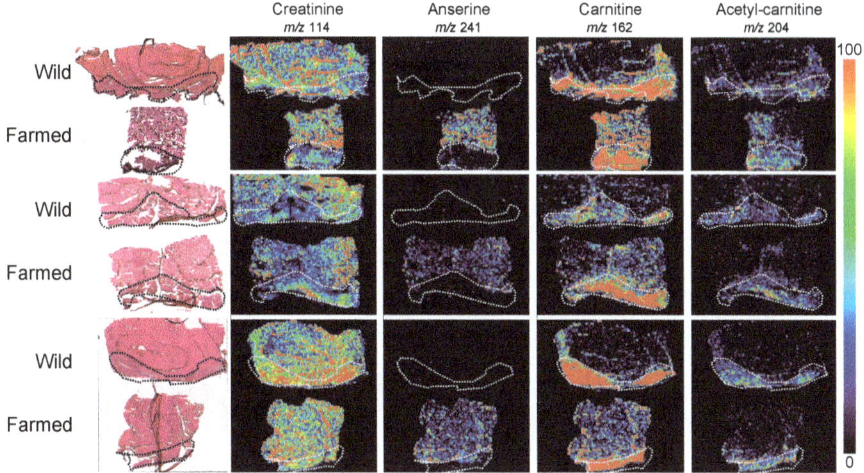

Figure 3. Matrix-assisted laser desorption/ionization mass spectrometry (MALDI-MS) imaging results of creatinine, anserine, carnitine, and acetyl carnitine distributions in wild and farmed red sea bream. Dotted outlines show the red muscle regions. Modified and reprinted with permission from [46].

Since the discovery of bovine spongiform encephalopathy, consumers have been faced with doubts about the safety of beef; thus, the geographical origin of meat is very important information that will influence consumer choice. Gene analysis and stable isotope analysis are applied to trace a beef sample to its geographic origin, but time-consuming procedures have made it difficult to distinguish individuals of the same breed according to location. Therefore, we applied MALDI-MS imaging to identify molecular markers of geographical origin based on principal component analysis in three independent beef tissue samples, which exhibited different geographical origins, and some key molecules were identified that could be used to classify their origin [37].

6. Recent Developments and Future Perspectives of Mass Spectrometry Imaging

Time-of-flight (TOF) technology is widely used for MS imaging, as this can effectively separate ionized accelerated molecules according to their mass-to-charge ratio (*m/z*). However, the mass resolution of TOF-MS is still not sufficient for a wide detection range since the single image is reconstructed at a resolution of ±0.1–0.2 Da, which might result in the overlap of multiple molecules within one constructed image. However, recent technological developments succeeded in achieving a higher mass resolution of approximately ±0.01 Da using Fourier-transform ion cyclotron resonance (FT-ICR) MS. For example, TOF-MS could not reveal the characteristic localization of thyroxine hormone (T3) in amphibian larvae owing to multiple mixed molecules including phospholipids in the vicinity of the molecular weight of the target. However, when the same sample was analyzed with FT-ICR MS, T3 and T4-specific peaks were observed (Figure 4) [77].

To obtain high spatial resolution, new matrix coating methods have also been established. Shuai et al. developed an electric field-assisted scanning-spraying matrix coating system to deposit the matrix on tissue with crystal sizes of <10 μm [78]. This system can easily generate homogenous and small matrix crystals on the tissue. It also enhanced the detection and imaging quality of tissue small molecule metabolites (<500 Da).

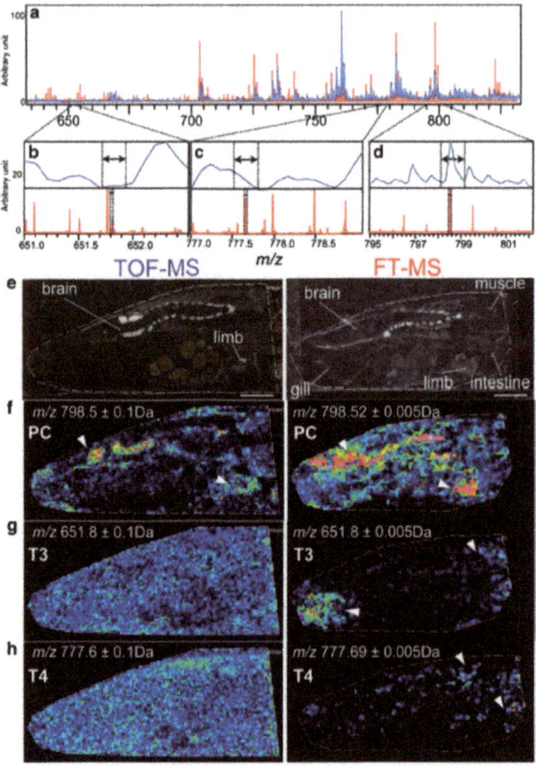

Figure 4. Time-of-flight mass spectrometry (TOF-MS) and Fourier-transform ion cyclotron resonance (FT-ICR) MS imaging results. Mass resolution was substantially higher in the FT-ICR mass spectrum, allowing for the visualizing of the specific localization of T3 and T4 in amphibian larvae. Modified and reprinted with permission from [77]. Mass spectra of tissue sections (**a–d**). Optical images of the tissue sections (**e**) and ion images of phosphatidylcholine, T3, and T4, respectively (**f–h**).

Recently, new ionization technology for higher ionization efficiency has been reported. Laser ablation DART imaging mass spectrometry was performed to detect the spatial distributions of highly volatile compounds in cross sections of *Coffea arabica* beans [79]. Moreover, a novel MS-based imaging platform was developed by integrating a new subatmospheric pressure MALDI source. This apparatus could perform in situ *N*-glycan imaging analysis with high resolution [80]. In 2019, Niehaus et al. developed a new ion source for transmission-mode geometry MALDI-MS imaging [81]. It can provide molecular information with a pixel size of 1 µm and smaller. This method could thus be a valuable new tool for research in cell biology.

MALDI-MS imaging generates an enormous amount of data. Therefore, diversification of analysis methods such as cloud-based analysis (e.g., SCiLS, https://scils.de/) will be needed in the near future to handle such big data for convenient interpretation. These data platforms could also perform multivariate statistical analyses, which could lead to meaningful data interpretation for researchers. In fact, datasets of unlimited size can be visualized and multivariate statistical analyses can be performed for detailed interpretation [82,83]. The data format imzML allows for the flexible and efficient exchange of MS imaging data between different instruments and data analysis software [84]. A number of software tools are available and many more are being adapted to imzML. (https://ms-imaging.org/wp/imzml/)

7. Conclusions

MALDI-MS imaging is a valuable tool to visualize food compounds and identify not only the nutritional content but also the geographical origin of the food for improved traceability, food safety, and breed enhancement, among other applications. We anticipate that MALDI-MS imaging will be used extensively in the food industry in the near future. However, certain challenges of this technology will need to be overcome, including the limited detection of molecules present at low concentrations or ionization efficiency. Therefore, further improvements to the method and/or new developments in the equipment should be a research focus to enable the sensitive detection of these molecules.

Author Contributions: Conceiving the study, M.M. and N.G.-I.; writing the manuscript, M.M., T.S., K.K., T.M. and N.G.-I.; all authors reviewed the manuscript.

Funding: The Naito Foundation (N. GI), JSPS Kakenhi Grant Number JP19K20161 (N. GI).

Conflicts of Interest: The authors declare no conflicts of interest.

References

1. Christou, C.; Poulli, S.; Yiannopoulos, E.; Agapiou, A. GC-MS analysis of D-pinitol in carob: Syrup and fruit (flesh and seed). *J. Chromatogr. B Anal. Technol. Biomed. Life Sci.* **2019**, *1116*, 60–64. [CrossRef] [PubMed]

2. Nimbalkar, M.S.; Pai, S.R.; Pawar, N.V.; Oulkar, D.; Dixit, G.B. Free amino acid profiling in grain amaranth using LC-MS/MS. *Food Chem.* **2012**, *134*, 2565–2569. [CrossRef] [PubMed]

3. Zdunic, G.; Godevac, D.; Savikin, K.; Krivokuca, D.; Mihailovic, M.; Przic, Z.; Markovic, N. Grape seed polyphenols and fatty acids of autochthonous Prokupac vine variety from Serbia. *Chem. Biodivers.* **2019**, *16*, e1900053. [CrossRef] [PubMed]

4. Gonzalez-Martin, M.I.; Revilla, M.I.; Vivar-Quintana, A.M.; Betances Salcedo, E.V. Pesticide residues in propolis from Spain and Chile. An. approach using near infrared spectroscopy. *Talanta* **2017**, *165*, 533–539. [CrossRef]

5. Ciampa, A.; Dell'Abate, M.T.; Masetti, O.; Valentini, M.; Sequi, P. Seasonal chemical–physical changes of PGI Pachino cherry tomatoes detected by magnetic resonance imaging (MRI). *Food Chem.* **2010**, *122*, 1253–1260. [CrossRef]

6. Harada, T.; Yuba-Kubo, A.; Sugiura, Y.; Zaima, N.; Hayasaka, T.; Goto-Inoue, N.; Wakui, M.; Suematsu, M.; Takeshita, K.; Ogawa, K.; et al. Visualization of volatile substances in different organelles with an atmospheric-pressure mass microscope. *Anal. Chem.* **2009**, *81*, 9153–9157. [CrossRef]

7. Burrell, M.; Earnshaw, C.; Clench, M. Imaging matrix assisted laser desorption ionization mass spectrometry: A technique to map plant metabolites within tissues at high spatial resolution. *J. Exp. Bot.* **2007**, *58*, 757–763. [CrossRef]

8. Mullen, A.K.; Clench, M.R.; Crosland, S.; Sharples, K.R. Determination of agrochemical compounds in soya plants by imaging matrix-assisted laser desorption/ionisation mass spectrometry. *Rapid Commun. Mass Spectrom.* **2005**, *19*, 2507–2516. [CrossRef]

9. Enomoto, H.; Takeda, S.; Hatta, H.; Zaima, N. Tissue-specific distribution of sphingomyelin species in pork chop revealed by matrix-assisted laser desorption/ionization-imaging mass spectrometry. *J. Food Sci.* **2019**, *84*, 1758–1763. [CrossRef]

10. Horn, P.J.; Silva, J.E.; Anderson, D.; Fuchs, J.; Borisjuk, L.; Nazarenus, T.J.; Shulaev, V.; Cahoon, E.B.; Chapman, K.D. Imaging heterogeneity of membrane and storage lipids in transgenic Camelina sativa seeds with altered fatty acid profiles. *Plant. J.* **2013**, *76*, 138–150.

11. Fuchs, B.; Bischoff, A.; Suss, R.; Teuber, K.; Schurenberg, M.; Suckau, D.; Schiller, J. Phosphatidylcholines and -ethanolamines can be easily mistaken in phospholipid mixtures: A negative ion MALDI-TOF MS study with 9-aminoacridine as matrix and egg yolk as selected example. *Anal. Bioanal. Chem.* **2009**, *395*, 2479–2487. [CrossRef] [PubMed]

12. Horn, P.J.; James, C.N.; Gidda, S.K.; Kilaru, A.; Dyer, J.M.; Mullen, R.T.; Ohlrogge, J.B.; Chapman, K.D. Identification of a new class of lipid droplet-associated proteins in plants. *Plant Physiol.* **2013**, *162*, 1926–1936. [CrossRef] [PubMed]

13. Chansela, P.; Goto-Inoue, N.; Zaima, N.; Sroyraya, M.; Sobhon, P.; Setou, M. Visualization of neuropeptides in paraffin-embedded tissue sections of the central nervous system in the decapod crustacean, Penaeus monodon, by imaging mass spectrometry. *Peptides* **2012**, *34*, 10–18. [CrossRef] [PubMed]

14. Groseclose, M.R.; Andersson, M.; Hardesty, W.M.; Caprioli, R.M. Identification of proteins directly from tissue: In situ tryptic digestions coupled with imaging mass spectrometry. *J. Mass Spectrom.* **2007**, *42*, 254–262. [CrossRef]

15. Rubakhin, S.S.; Li, L.; Moroz, T.P.; Sweedler, J.V. Characterization of the Aplysia californica cerebral ganglion F cluster. *J Neurophysiol.* **1999**, *81*, 1251–1260. [CrossRef]

16. Caprioli, R.M.; Farmer, T.B.; Gile, J. Molecular imaging of biological samples: Localization of peptides and proteins using MALDI-TOF MS. *Anal. Chem.* **1997**, *69*, 4751–4760. [CrossRef]

17. Stoeckli, M.; Chaurand, P.; Hallahan, D.E.; Caprioli, R.M. Imaging mass spectrometry: A new technology for the analysis of protein expression in mammalian tissues. *Nat. Med.* **2001**, *7*, 493–496. [CrossRef]

18. Aubagnac, J.L.; Enjalbal, C.; Drouot, C.; Combarieu, R.; Martinez, J. Imaging time-of-flight secondary ion mass spectrometry of solid-phase peptide syntheses. *J. Mass Spectrom.* **1999**, *34*, 749–754. [CrossRef]

19. Trim, P.J.; Henson, C.M.; Avery, J.L.; McEwen, A.; Snel, M.F.; Claude, E.; Marshall, P.S.; West, A.; Princivalle, A.P.; Clench, M.R. Matrix-assisted laser desorption/ionization-ion mobility separation-mass spectrometry imaging of vinblastine in whole body tissue sections. *Anal. Chem.* **2008**, *80*, 8628–8634. [CrossRef]

20. Goodwin, R.J.; Scullion, P.; Macintyre, L.; Watson, D.G.; Pitt, A.R. Use of a solvent-free dry matrix coating for quantitative matrix-assisted laser desorption ionization imaging of 4-bromophenyl-1,4-diazabicyclo(3.2.2)nonane-4-carboxylate in rat brain and quantitative analysis of the drug from laser microdissected tissue regions. *Anal. Chem.* **2010**, *82*, 3868–3873.

21. Chaurand, P.; Rahman, M.A.; Hunt, T.; Mobley, J.A.; Gu, G.; Latham, J.C.; Caprioli, R.M.; Kasper, S. Monitoring mouse prostate development by profiling and imaging mass spectrometry. *Mol. Cell Proteom.* **2008**, *7*, 411–423. [CrossRef]

22. Yoshimura, Y.; Goto-Inoue, N.; Moriyama, T.; Zaima, N. Significant advancement of mass spectrometry imaging for food chemistry. *Food Chem.* **2016**, *210*, 200–211. [CrossRef]

23. Goto-Inoue, N.; Setou, M.; Zaima, N. Visualization of spatial distribution of gamma-aminobutyric acid in eggplant (Solanum melongena) by matrix-assisted laser desorption/ionization imaging mass spectrometry. *Anal. Sci.* **2010**, *26*, 821–825. [CrossRef]

24. Enomoto, H.; Sato, K.; Miyamoto, K.; Ohtsuka, A.; Yamane, H. Distribution analysis of anthocyanins, sugars, and organic acids in strawberry fruits using matrix-assisted laser desorption/ionization-imaging mass spectrometry. *J. Agric. Food Chem.* **2018**, *66*, 4958–4965. [CrossRef]

25. Taira, S.; Shimma, S.; Osaka, I.; Kaneko, D.; Ichiyanagi, Y.; Ikeda, R.; Konishi-Kawamura, Y.; Zhu, S.; Tsuneyama, K.; Komatsu, K. Mass spectrometry imaging of the capsaicin localization in the capsicum fruits. *Int. J. Biotechnol. Wellness Ind.* **2012**, *1*, 61–65. [CrossRef]

26. Agar, N.Y.; Golby, A.J.; Ligon, K.L.; Norton, I.; Mohan, V.; Wiseman, J.M.; Tannenbaum, A.; Jolesz, F.A. Development of stereotactic mass spectrometry for brain tumor surgery. *Neurosurgery* **2011**, *68*, 280–289. [CrossRef]

27. Eberlin, L.S.; Dill, A.L.; Costa, A.B.; Ifa, D.R.; Cheng, L.; Masterson, T.; Koch, M.; Ratliff, T.L.; Cooks, R.G. Cholesterol sulfate imaging in human prostate cancer tissue by desorption electrospray ionization mass spectrometry. *Anal. Chem.* **2010**, *82*, 3430–3434. [CrossRef]

28. Masaki, N.; Ishizaki, I.; Hayasaka, T.; Fisher, G.L.; Sanada, N.; Yokota, H.; Setou, M. Three-Dimensional image of cleavage bodies in nuclei is configured using gas cluster ion beam with time-of-flight secondary ion. mass spectrometry. *Sci. Rep.* **2015**, *5*, 10000. [CrossRef]

29. Takats, Z.; Wiseman, J.M.; Gologan, B.; Cooks, R.G. Mass spectrometry sampling under ambient conditions with desorption electrospray ionization. *Science* **2004**, *306*, 471–473. [CrossRef]

30. Eberlin, L.S.; Ferreira, C.R.; Dill, A.L.; Ifa, D.R.; Cheng, L.; Cooks, R.G. Nondestructive, histologically compatible tissue imaging by desorption electrospray ionization mass spectrometry. *ChemBioChem* **2011**, *12*, 2129–2132. [CrossRef]

31. Enomoto, H.; Sensu, T.; Sato, K.; Sato, F.; Paxton, T.; Yumoto, E.; Miyamoto, K.; Asahina, M.; Yokota, T.; Yamane, H. Visualisation of abscisic acid and 12-oxo-phytodienoic acid in immature Phaseolus vulgaris L. seeds using desorption electrospray ionisation-imaging mass spectrometry. *Sci. Rep.* **2017**, *7*, 42977. [CrossRef] [PubMed]

32. Liao, Y.; Fu, X.; Zhou, H.; Rao, W.; Zeng, L.; Yang, Z. Visualized analysis of within-tissue spatial distribution of specialized metabolites in tea (*Camellia sinensis*) using desorption electrospray ionization imaging mass spectrometry. *Food Chem.* **2019**, *292*, 204–210. [CrossRef] [PubMed]

33. Li, Y.; Shrestha, B.; Vertes, A. Atmospheric pressure molecular imaging by infrared MALDI mass spectrometry. *Anal. Chem.* **2007**, *79*, 523–532. [CrossRef] [PubMed]

34. Robinson, S.; Warburton, K.; Seymour, M.; Clench, M.; Thomas-Oates, J. Localization of water-soluble carbohydrates in wheat stems using imaging matrix-assisted laser desorption ionization mass spectrometry. *N. Phytol.* **2007**, *173*, 438–444. [CrossRef] [PubMed]

35. Sroyraya, M.; Goto-Inoue, N.; Zaima, N.; Hayasaka, T.; Chansela, P.; Tanasawet, S.; Shrivas, K.; Sobhon, P.; Setou, M. Visualization of biomolecules in the eyestalk of the blue swimming crab, Portunus pelagicus, by imaging mass spectrometry using the atmospheric-pressure mass microscope. *Surf. Interface Anal.* **2010**, *42*, 1589–1592. [CrossRef]

36. Zaima, N.; Goto-Inoue, N.; Hayasaka, T.; Setou, M. Application of imaging mass spectrometry for the analysis of *Oryza sativa* rice. *Rapid Commun. Mass Spectrom.* **2010**, *24*, 2723–2729. [CrossRef]

37. Zaima, N.; Goto-Inoue, N.; Hayasaka, T.; Enomoto, H.; Setou, M. Authenticity assessment of beef origin by principal component analysis of matrix-assisted laser desorption/ionization mass spectrometric data. *Anal. Bioanal. Chem.* **2011**, *400*, 1865–1871. [CrossRef]

38. Yoshimura, Y.; Zaima, N.; Moriyama, T.; Kawamura, Y. Different localization patterns of anthocyanin species in the pericarp of black rice revealed by imaging mass spectrometry. *PLoS ONE* **2012**, *7*, e31285. [CrossRef]

39. Hashizaki, R.; Komori, H.; Kazuma, K.; Konno, K.; Kawabata, K.; Kaneko, D.; Katano, H.; Taira, S. Localization analysis of natural toxin of *Solanum tuberosum* L. via mass spectrometric imaging. *Int. J. Biotechnol. Well. Ind.* **2016**, *5*, 1–5.

40. Velickovic, D.; Saulnier, L.; Lhomme, M.; Damond, A.; Guillon, F.; Rogniaux, H. Mass Spectrometric Imaging of Wheat (*Triticum* spp.) and Barley (*Hordeum vulgare* L.) Cultivars: Distribution of major cell wall polysaccharides according to their main structural features. *J. Agric. Food Chem.* **2016**, *64*, 6249–6256. [CrossRef]

41. Nakamura, J.; Morikawa-Ichinose, T.; Fujimura, Y.; Hayakawa, E.; Takahashi, K.; Ishii, T.; Miura, D.; Wariishi, H. Spatially resolved metabolic distribution for unraveling the physiological change and responses in tomato fruit using matrix-assisted laser desorption/ionization-mass spectrometry imaging (MALDI-MSI). *Anal. Bioanal. Chem.* **2017**, *409*, 1697–1706. [CrossRef] [PubMed]

42. Shiono, K.; Hashizaki, R.; Nakanishi, T.; Sakai, T.; Yamamoto, T.; Ogata, K.; Harada, K.I.; Ohtani, H.; Katano, H.; Taira, S. Multi-imaging of cytokinin and abscisic acid on the roots of rice (*Oryza sativa*) using matrix-assisted laser desorption/ionization mass spectrometry. *J. Agric. Food Chem.* **2017**, *65*, 7624–7628. [CrossRef] [PubMed]

43. Ekelof, M.; McMurtrie, E.K.; Nazari, M.; Johanningsmeier, S.D.; Muddiman, D.C. Direct analysis of triterpenes from high.-salt fermented cucumbers using Infrared Matrix-Assisted Laser Desorption Electrospray Ionization (IR-MALDESI). *J. Am. Soc. Mass Spectrom.* **2017**, *28*, 370–375. [CrossRef] [PubMed]

44. Feenstra, A.D.; Alexander, L.E.; Song, Z.; Korte, A.R.; Yandeau-Nelson, M.D.; Nikolau, B.J.; Lee, Y.J. Spatial mapping and profiling of metabolite distributions during germination. *Plant Physiol* **2017**, *174*, 2532–2548. [CrossRef] [PubMed]

45. Woodfield, H.K.; Sturtevant, D.; Borisjuk, L.; Munz, E.; Guschina, I.A.; Chapman, K.; Harwood, J.L. Spatial and temporal mapping of key lipid species in *Brassica napus* seeds. *Plant Physiol.* **2017**, *173*, 1998–2009. [CrossRef] [PubMed]

46. Goto-Inoue, N.; Sato, T.; Morisasa, M.; Igarashi, Y.; Mori, T. Characterization of Metabolite compositions in wild and farmed red sea bream (*Pagrus major*) using mass spectrometry imaging. *J. Agric. Food Chem.* **2019**, *67*, 7197–7203. [CrossRef] [PubMed]

47. Fanuel, M.; Ropartz, D.; Guillon, F.; Saulnier, L.; Rogniaux, H. Distribution of cell wall hemicelluloses in the wheat grain endosperm: A 3D perspective. *Planta* **2018**, *248*, 1505–1513. [CrossRef]

48. Resetar Maslov, D.; Svirkova, A.; Allmaier, G.; Marchetti-Deschamann, M.; Kraljevic Pavelic, S. Optimization of MALDI-TOF mass spectrometry imaging for the visualization and comparison of peptide distributions in dry-cured ham muscle fibers. *Food Chem.* **2019**, *283*, 275–286. [CrossRef]

49. Horikawa, K.; Hirama, T.; Shimura, H.; Jitsuyama, Y.; Suzuki, T. Visualization of soluble carbohydrate distribution in apple fruit flesh utilizing MALDI-TOF MS imaging. *Plant Sci.* **2019**, *278*, 107–112. [CrossRef]

50. Bednarz, H.; Roloff, N.; Niehaus, K. Mass spectrometry imaging of the spatial and temporal localization of alkaloids in Nightshades. *J. Agric. Food Chem.* **2019**. [CrossRef]

51. Jehl, B.; Bauer, R.; Dorge, A.; Rick, R. The use of propane/isopentane mixtures for rapid freezing of biological specimens. *J. Microsc.* **1981**, *123*, 307–309. [CrossRef] [PubMed]

52. Ly, A.; Longuespee, R.; Casadonte, R.; Wandernoth, P.; Schwamborn, K.; Bollwein, C.; Marsching, C.; Kriegsmann, K.; Hopf, C.; Weichert, W.; et al. Site-to-site reproducibility and spatial resolution in MALDI-MSI of peptides from formalin-fixed paraffin-embedded samples. *Proteom. Clin. Appl.* **2019**, *13*, e1800029. [CrossRef] [PubMed]

53. Schwartz, S.A.; Reyzer, M.L.; Caprioli, R.M. Direct tissue analysis using matrix-assisted laser desorption/ionization mass spectrometry: Practical aspects of sample preparation. *J. Mass Spectrom.* **2003**, *38*, 699–708. [CrossRef] [PubMed]

54. Stoeckli, M.; Staab, D.; Schweitzer, A. Compound and metabolite distribution measured by MALDI mass spectrometric imaging in whole-body tissue sections. *Int. J. Mass Spectrom.* **2007**, *260*, 195–202. [CrossRef]

55. Khatib-Shahidi, S.; Andersson, M.; Herman, J.L.; Gillespie, T.A.; Caprioli, R.M. Direct molecular analysis of whole-body animal tissue sections by imaging MALDI mass spectrometry. *Anal. Chem.* **2006**, *78*, 6448–6456. [CrossRef]

56. Kawamoto, T. Use of a new adhesive film for the preparation of multi-purpose fresh-frozen sections from hard tissues, whole-animals, insects and plants. *Arch. Histol. Cytol.* **2003**, *66*, 123–143. [CrossRef]

57. Fujino, Y.; Minamizaki, T.; Yoshioka, H.; Okada, M.; Yoshiko, Y. Imaging and mapping of mouse bone using MALDI-imaging mass spectrometry. *Bone Rep.* **2016**, *5*, 280–285. [CrossRef]

58. Seeley, E.H.; Wilson, K.J.; Yankeelov, T.E.; Johnson, R.W.; Gore, J.C.; Caprioli, R.M.; Matrisian, L.M.; Sterling, J.A. Co-registration of multi-modality imaging allows for comprehensive analysis of tumor-induced bone disease. *Bone* **2014**, *61*, 208–216. [CrossRef]

59. Nakabayashi, R.; Hashimoto, K.; Toyooka, K.; Saito, K. Keeping the shape of plant tissue for visualizing metabolite features in segmentation and correlation analysis of imaging mass spectrometry in Asparagus officinalis. *Metabolomics* **2019**, *15*, 24. [CrossRef]

60. Enthaler, B.; Bussmann, T.; Pruns, J.K.; Rapp, C.; Fischer, M.; Vietzke, J.P. Influence of various on-tissue washing procedures on the entire protein quantity and the quality of matrix-assisted laser desorption/ionization spectra. *Rapid Commun. Mass Spectrom.* **2013**, *27*, 878–884. [CrossRef]

61. Shimma, S.; Furuta, M.; Ichimura, K.; Yoshida, Y.; Setou, M. Direct MS/MS analysis in mammalian tissue sections using MALDI-QIT-TOFMS and chemical inkjet technology. *Surf. Interface Anal.* **2006**, *38*, 1712–1714. [CrossRef]

62. Heijs, B.; Carreira, R.J.; Tolner, E.A.; de Ru, A.H.; van den Maagdenberg, A.M.; van Veelen, P.A.; McDonnell, L.A. Comprehensive analysis of the mouse brain proteome sampled in mass spectrometry imaging. *Anal. Chem.* **2015**, *87*, 1867–1875. [CrossRef] [PubMed]

63. Enthaler, B.; Trusch, M.; Fischer, M.; Rapp, C.; Pruns, J.K.; Vietzke, J.P. MALDI imaging in human skin tissue sections: Focus on various matrices and enzymes. *Anal. Bioanal. Chem.* **2013**, *405*, 1159–1170. [CrossRef] [PubMed]

64. Powers, T.W.; Jones, E.E.; Betesh, L.R.; Romano, P.R.; Gao, P.; Copland, J.A.; Mehta, A.S.; Drake, R.R. Matrix assisted laser desorption ionization imaging mass spectrometry workflow for spatial profiling analysis of N-linked glycan expression in tissues. *Anal. Chem.* **2013**, *85*, 9799–9806. [CrossRef]

65. Pinho, S.S.; Reis, C.A. Glycosylation in cancer: Mechanisms and clinical implications. *Nat. Rev. Cancer* **2015**, *15*, 540–555. [CrossRef]

66. Adamczyk, B.; Tharmalingam, T.; Rudd, P.M. Glycans as cancer biomarkers. *Biochim. Biophys. Acta* **2012**, *1820*, 1347–1353. [CrossRef]

67. Powers, T.W.; Neely, B.A.; Shao, Y.; Tang, H.; Troyer, D.A.; Mehta, A.S.; Haab, B.B.; Drake, R.R. MALDI imaging mass spectrometry profiling of N-glycans in formalin-fixed paraffin embedded clinical tissue blocks and tissue microarrays. *PLoS ONE* **2014**, *9*, e106255. [CrossRef]

68. Bagley, M.C.; Stepanova, A.N.; Ekelof, M.; Alonso, J.M.; Muddiman, D.C. Development of a relative quantification method for IR-MALDESI mass spectrometry imaging of Arabidopsis root seedlings. *Rapid Commun. Mass Spectrom.* **2019**. [CrossRef]

69. Shariatgorji, M.; Nilsson, A.; Goodwin, R.J.; Kallback, P.; Schintu, N.; Zhang, X.; Crossman, A.R.; Bezard, E.; Svenningsson, P.; Andren, P.E. Direct targeted quantitative molecular imaging of neurotransmitters in brain tissue sections. *Neuron* **2014**, *84*, 697–707. [CrossRef]

70. Karas, M.; Hillenkamp, F. Laser desorption ionization of proteins with molecular masses exceeding 10,000 daltons. *Anal. Chem.* **1988**, *60*, 2299–2301. [CrossRef]

71. Tanaka, K.; Waki, H.; Ido, Y.; Akita, S.; Yoshida, Y.; Yoshida, T.; Matsuo, T. Protein and polymer analyses up to m/z 100 000 by laser ionization time-of-flight mass spectrometry. *Rapid Commun. Mass Spectrom.* **1988**, *2*, 151–153. [CrossRef]

72. Chughtai, K.; Heeren, R.M. Mass spectrometric imaging for biomedical tissue analysis. *Chem. Rev.* **2010**, *110*, 3237–3277. [CrossRef] [PubMed]

73. Kaletas, B.K.; van der Wiel, I.M.; Stauber, J.; Lennard, J.D.; Guzel, C.; Kros, J.M.; Luider, T.M.; Heeren, R.M. Sample preparation issues for tissue imaging by imaging MS. *Proteomics* **2009**, *9*, 2622–2633. [CrossRef] [PubMed]

74. Agar, N.Y.; Yang, H.W.; Carroll, R.S.; Black, P.M.; Agar, J.N. Matrix solution fixation: Histology-compatible tissue preparation for MALDI mass spectrometry imaging. *Anal. Chem.* **2007**, *79*, 7416–7423. [CrossRef]

75. Gemperline, E.; Rawson, S.; Li, L. Optimization and comparison of multiple MALDI matrix application methods for small molecule mass spectrometric imaging. *Anal. Chem.* **2014**, *86*, 10030–10035. [CrossRef]

76. Shimma, S.; Sugiura, Y. effective sample preparations in imaging mass spectrometry. *Mass Spectrom.* **2014**, *3*, S0029. [CrossRef]

77. Goto-Inoue, N.; Sato, T.; Morisasa, M.; Kashiwagi, A.; Kashiwagi, K.; Sugiura, Y.; Sugiyama, E.; Suematsu, M.; Mori, T. Utilizing mass spectrometry imaging to map the thyroid hormones triiodothyronine and thyroxine in *Xenopus tropicalis* tadpoles. *Anal. Bioanal. Chem.* **2018**, *410*, 1333–1340. [CrossRef]

78. Guo, S.; Wang, Y.; Zhou, D.; Li, Z. Electric field-assisted matrix coating method enhances the detection of small molecule metabolites for mass spectrometry imaging. *Anal. Chem.* **2015**, *87*, 5860–5865. [CrossRef]

79. Fowble, K.L.; Okuda, K.; Cody, R.B.; Musah, R.A. Spatial distributions of furan and 5-hydroxymethylfurfural in unroasted and roasted *Coffea arabica* beans. *Food Res. Int.* **2019**, *119*, 725–732. [CrossRef]

80. Shi, Y.; Li, Z.; Felder, M.A.; Yu, Q.; Shi, X.; Peng, Y.; Cao, Q.; Wang, B.; Puglielli, L.; Patankar, M.S.; et al. Mass spectrometry imaging of N-glycans from formalin-fixed paraffin-embedded tissue sections using a novel subatmospheric pressure ionization source. *Anal. Chem.* **2019**, *91*, 12942–12947. [CrossRef]

81. Niehaus, M.; Soltwisch, J.; Belov, M.E.; Dreisewerd, K. Transmission-mode MALDI-2 mass spectrometry imaging of cells and tissues at subcellular resolution. *Nat. Methods* **2019**, *16*, 925–931. [CrossRef] [PubMed]

82. Klein, O.; Strohschein, K.; Nebrich, G.; Fuchs, M.; Thiele, H.; Giavalisco, P.; Duda, G.N.; Winkler, T.; Kobarg, J.H.; Trede, D.; et al. Unraveling local tissue changes within severely injured skeletal muscles in response to MSC-based intervention using MALDI Imaging mass spectrometry. *Sci. Rep.* **2018**, *8*, 12677. [CrossRef] [PubMed]

83. Paine, M.R.L.; Liu, J.; Huang, D.; Ellis, S.R.; Trede, D.; Kobarg, J.H.; Heeren, R.M.A.; Fernandez, F.M.; MacDonald, T.J. Three-dimensional mass spectrometry imaging identifies lipid markers of medulloblastoma metastasis. *Sci. Rep.* **2019**, *9*, 2205. [CrossRef] [PubMed]

84. Schramm, T.; Hester, Z.; Klinkert, I.; Both, J.P.; Heeren, R.M.A.; Brunelle, A.; Laprevote, O.; Desbenoit, N.; Robbe, M.F.; Stoeckli, M.; et al. ImzML—A common data format for the flexible exchange and processing of mass spectrometry imaging data. *J. Proteom.* **2012**, *75*, 5106–5110. [CrossRef] [PubMed]

Article

Valorization of Prickly Pear Juice Geographical Origin Based on Mineral and Volatile Compound Contents Using LDA

Vassilios K. Karabagias [1], Ioannis K. Karabagias [1,*], Artemis Louppis [2], Anastasia Badeka [1], Michael G. Kontominas [1] and Chara Papastephanou [2]

[1] Laboratory of Food Chemistry Department of Chemistry University of Ioannina, 45110 Ioannina, Greece; vkarambagias@gmail.com (V.K.K.); abadeka@uoi.gr (A.B.); mkontomi@cc.uoi.gr (M.G.K.)

[2] cp Foodlab Ltd, Polifonti 25, Strovolos, Nicosia 2047, Cyprus; artemislouppis@gmail.com (A.L.); foodlab@cytanet.com.cy (C.P.)

* Correspondence: ikaraba@cc.uoi.gr; Tel.: +30-697-828-6866

Received: 23 March 2019; Accepted: 11 April 2019; Published: 15 April 2019

Abstract: In the present work the mineral content and volatile profile of prickly pear juice prepared from wild cultivars was investigated. Fruits used in the study originated from three areas of the Peloponnese Peninsula. Twenty-five macro- and micro-minerals (K, Na, P, Ca, Mg, Al, B, Ba, Be, Co, Cr, Cu, Fe, Li, Mn, Mo, Ni, Sb, Se, Si, Sn, Ti, Tl, V, Zn) were determined using inductively coupled plasma atomic emission spectroscopy (ICP-OES). Furthermore, analysis of the mineral content of soil samples with ICP-OES showed a perfect correlation with those of fruit juices. Volatile compounds (alcohols, aldehydes, hydrocarbons, terpenoids, and others) were identified using an optimized headspace solid phase microextraction coupled to gas chromatography mass spectrometry (HS-SPME/GC-MS) method. Multivariate analysis showed significant differences ($p < 0.05$) among the investigated parameters with respect to juice geographical origin. Prickly pear juice samples were classified according to geographical origin by 85.7% and 88.9% using 7 minerals and 21 volatile compounds, respectively.

Keywords: prickly pear; juice; minerals; aroma; authentication; chemometrics

---

1. Introduction

Opuntia ficus indica or the Indian fig opuntia is a fruit that belongs to the cactus family, Cactaceae [1]. The fruit originated from North and South America; however, it was also introduced in Europe and grows in regions with a suitable climate, such as the south of France, southern Italy (Sardinia, Sicily, etc.), Bulgaria, southern Portugal (Madeira), Spain (Andalusia), Albania, Cyprus, and Greece. Some typical regions in Greece where the fruit grows well are Peloponnese, the Ionian Islands, and Crete.

From a nutritional point of view, prickly pear fruit (portions of 100g) provides the human body with 41 calories and is composed of water (ca. 88%), carbohydrates (ca. 10%), and smaller amounts of fat, protein, vitamin C, and minerals [2]. Considerable amounts of phytochemicals (betalain, betanin, indicaxanthin, gallic acid, vanillic acid, catechins, etc.) have also been reported [3,4]. The fruit has a long term history in the Mexican culture as a natural healer of wounds and inflammation of the digestive and urinary tracts [5].

Fruit juices may comprise a delicious beverage with a characteristic flavor and taste, and may be a health companion in daily diets, due to the documented health benefits upon regular consumption. Some typical components that are responsible for the beneficial health effects of fruit juices include polyphenols, carotenoids, vitamins, minerals and trace elements [6,7]. Yet, a unique aroma may be the primary criterion for acceptance of a product among consumers. In that sense, when sensory and nutritional characteristics are well combined, these attributes may result in the development of a new product.

Some previous studies in the literature deal with the characterization of apple, orange, pear, peach, apricot, blueberry, cranberry, plum, lemon, and cherry juices by means of volatile compounds, polyphenols, spectra profiling, mineral and trace element analysis, and sensory metrics determination, using solid phase micro-extraction coupled to gas chromatography mass spectrometry (SPME-GC/MS), high performance liquid chromatography (HPLC), Fourier transform infrared spectroscopy (FTIR), nuclear magnetic resonance (1H-NMR), inductively coupled plasma atomic (or optical) emission spectrometry (ICP-AES or ICP-OES), inductively coupled plasma mass spectrometry (ICP-MS), instrumental neutronic activation analysis (INAA) and the electronic tongue [8–14].

As a matter of fact, application of innovative statistical approaches on the aforementioned chemical or sensory markers, namely principal component analysis (PCA), linear discriminant analysis (LDA), soft independent modeling of class analogies (SIMCA), hierarchical cluster analysis (HCA), k-nearest neighbor analysis (KNN), and partial least square-discriminant analysis (PLS-DA), has led to the development of accurate models for fruit juice authentication [7].

However, data involving the mineral content and volatile profile of prickly pear juice prepared from Greek wild-grown prickly pear fruits has not been previously reported. In addition, correlation of the mineral content of soil samples with those of fruit juice samples has been scarcely reported.

Considering the above, the objectives of the present study were to: (i) investigate the mineral content of prickly pear juice and correlate it with the mineral content of soil, where fruits of the naturally grown wild cultivars were collected, (ii) determine the volatile compounds that are responsible for prickly pear juice aroma, and (iii) check whether the differences in mineral or volatile compound content may be used for prickly pear juice differentiation according to geographical origin using a supervised chemometric tool, such as linear discriminant analysis. Practical applications of the present study may be focused on two basic axes: (i) exploitation of prickly pear juice and (ii) creation of a data base of mineral and volatile compound content of prickly pear juice prepared from wild fruit cultivars, as these are the intermediates for the cultivation of hybrid cultivars often termed as "less wild ones".

2. Materials and Methods

2.1. Prickly Pear Juice Samples

Thirty-six batch juice samples were prepared by machine squeezing (Rohnson fruit squeezer, power of 1000 W) of ca. 60 kg of prickly pear fruit naturally grown (wild cultivar) in the regions of East Messinia (12 samples), West Messinia (12 samples), and Lakonia (12 samples). This procedure (batch sampling) was followed in order to: (i) eliminate any existing differences in the maturity of the fruit used since these originated from different cactus plants and (ii) reduce the experimental cost of the study. Prior to juice preparation, fruit were tentatively washed with tap water, dried, and manually peeled. Samples were stored in polyethylene terephthalate (PET) containers (volume of 500 mL) and maintained at −18 ± 1 °C until analyses.

2.2. Chemicals and Multi-Element Standard

The chemicals used in the study were of analytical grade. The standard solutions of each mineral were prepared by appropriate dilution with ultrapure water (Milli-Q, Millipore, Bedford, MA, USA), of a multi-element standard (100 mg/L) obtained from Merck (Darmstadt, Germany). Nitric acid suprapure 65%, used for the digestion of samples, was obtained from Merck (Darmstadt, Germany) [15].

2.3. Determination of Mineral Content in Soil Samples

The mineral content of soil samples was determined according to the method of the Association of Official Analytical Chemists (AOAC) [16], which refers to the determination of metals in solid wastes, by using an inductively coupled plasma atomic emission spectrometric method (ICP-OES). The analysis of each sample was carried out in triplicate ($n = 3$). Results were expressed as mg/L.

2.4. Determination of Mineral Content in Prickly Pear Juice Samples

Approximately 0.5 g of each juice sample or fruit were weighed in microwave cup, previously rinsed with a mixture of superoxide and water (1:1). The sample was then mineralized using 7 mL 65% HNO_3 (suprapure) and 2 mL of H_2O_2 20% (*v/v*) (Merck, Darmstadt, Germany). The digestion was accomplished by heating the mixture for 10 and 20 min at 200 °C, respectively, using a microwave digester (power of 1000 W). The obtained mixture was then sonicated and diluted to a final volume of 100 mL with ultrapure water before ICP-OES analysis. The analysis of each sample was carried out in triplicate ($n = 3$). Results were expressed as mg/kg.

2.5. ICP-OES Instrumentation and Method Analytical Characteristics

A Thermo Scientific IRIS Intrepid II XDL inductively coupled plasma-atomic emission spectrometer (Thermo Electron Corporation, Waltham, MA, USA) was used for the elemental analysis. The emission wavelength (nm) was: 309.3, 455.4, 455.4, 313.0, 393.3, 228.6, 267.7, 324.7, 259.9, 766.5, 670.8, 279.5, 257.6, 202.0, 589.0, 221.6, 178.3, 206.8, 196.0, 251.6, 190.0, 334.9, 190.8, 292.4, 213.9, for Al, Ba, B, Be, Ca, Co, Cr, Cu, Fe, K, Li, Mg, Mn, Mo, Na, Ni, P, Sb, Se, Si, Sn, Ti, Tl, V and Zn, respectively. The operational parameters for the instrument were in accordance with a previous work [17]. For the determination of each amount of mineral, calibration curves were prepared. These showed correlation coefficients (R^2) in the range of 0.9967–1.000 for soil and 0.997–1.000 for fruit juice samples, respectively (Supplementary Tables S1 and S2). Furthermore, the analytical method developed in the present study showed a satisfactory percent recovery for both soil and prickly pear juice samples at different spiking concentrations (six replicates) (Supplementary Tables S3 and S4). The limit of detection (LOD) and limit of quantification (LOQ) were estimated by spiking a blank sample (ultrapure water) three times with the standard mineral solution at low concentrations and the signal-to-noise ratio was determined. The LOD was defined as 3:1 and the LOQ as 10:1. The LOD and LOQ values for each mineral determined in soil and fruit juice samples are provided in the supplementary material (Supplementary Tables S5 and S6). Finally, the coefficient of variation for all minerals determined in prickly pear juice samples ($n = 36$) was ≤5.10% (Supplementary Table S7).

2.6. HS-SPME/GC-MS Analysis

2.6.1. Method Optimization

An optimization procedure was followed, in a preliminary experiment, in order to determine the most appropriate parameters for the extraction of volatile compounds from the headspace of prickly pear juice. These included: sample volume (5 and 10 mL), equilibrium time (10, 20, 25, and 30 min), sampling time (10 and 20 min), sample volume, extraction temperature (40, 42, and 45 °C), and salt (0, 20, and 30% *w/v*) addition [18,19]. Based on (i) the number of volatiles determined, (ii) the MS qualification results, (iii) the limited furan derivatives identified, and (iv) the spectra intensity along with the agreement in volatiles identified during the analysis of replicates, the optimum analysis conditions were found to be: 25 min equilibration time, 20 min sampling time, 5 mL sample volume, addition of salt (30% *w/v*), and 42 °C water bath temperature (Supplementary Table S8 and typical chromatograms (Figures S1–S8)).

2.6.2. Extraction of Volatile Compounds

The volatile compounds of prickly pear juice were extracted using a divinyl benzene/carboxen/ polydimethylsiloxane (DVB/CAR/PDMS) fiber 50/30 μm (Supelco, Bellefonte, PA, USA). The fiber was firstly conditioned, prior to use, according to the manufacturer's recommendations. Afterwards, the samples consisting of 5 mL of juice, 1.5 g NaCl (Merck, Darmstadt, Germany), and 50 μL of internal standard (benzophenone, 100 µg/mL, Sigma Aldrich, St. Louis, MO, USA), were placed in 20 mL screw-cap vials equipped with polytetrafluoroethylene (PTFE) septa. The vials were maintained at 42 °C in a water bath under continuous stirring at 800 rpm during the headspace extraction. The stirring procedure usually improves the extraction efficiency [19]. For this purpose, a magnetic stirrer (cross shaped and coated with PTFE) with a diameter of 10 mm (Semadeni, Ostermundigen–Bern, Switzerland) was placed inside the vials. Fruit juice samples were prepared daily prior the HS-SPME-GC/MS analysis and the fiber was cleaned, before each sample analysis, using the clean program method. The analysis of each sample was run in duplicate and results were averaged.

2.6.3. GC/MS Instrumentation and Analysis Conditions

Volatile compound analysis of prickly pear juice samples was carried out using an Agilent gas chromatograph (Agilent 7890A) equipped with an Agilent mass detector (Agilent 5975). The chromatographic separation was achieved using DB-5MS (cross linked 5% PH ME siloxane) capillary column (60 m × 320 µm i.d., × 1 µm film thickness). Helium was used as the carrier gas (purity of 99.999%) at a flow rate of 1.5 mL/min. The injector and MS-transfer line were maintained at 260 °C and 280 °C, respectively. For HS-SPME analysis, the initial oven temperature of 40 °C was increased to 168 °C at a rate of 4 °C/min (0 min hold), and finally increased at a rate of 10 °C/min to 260 °C (1 min hold). The acquisition mode was scan. Electron impact mass spectra were recorded in the 29–350 mass range. An electron ionization system was used with an ionization energy of 70 eV. Solvent delay was set at 5 min to avoid co-elution of ethanol, since the internal standard was dissolved in ethanol [19]. Finally, a split ratio of 2:1 was used in the analysis.

2.6.4. Identification of Volatile Compounds

The Wiley 7, NIST 2005 mass spectral library was used for the identification of volatile compounds of prickly pear juice. The linear retention indices were also calculated for each compound, using a mixture of n-alkanes (C8–C20) dissolved in n-hexane. The standard mixture of alkanes was purchased by Fluka (Leipzig, Germany). The calculation was carried out for components eluting between n-octane and n-eicosane. Only the volatile compounds that had >85% similarity with Wiley library were tentatively identified using the GC-MS spectra. Results were expressed as a concentration ($C_{analyte}$, µg/L) based on the peak area ratio of the isolated volatile compounds to that of the internal standard assuming a response factor equal to one for all the isolated compounds [18]. The internal standard (m/z = 182) used did not cause any co-elution problems, as it eluted as the final organic compound with no derivatives were identified.

2.7. Statistical Treatment of Data

Statistical treatment of data was performed using the SPSS 20.0 statistics software (IBM, Armonk, NY, USA). Comparison of average values (mineral content and semi-quantitative data of volatile compounds) was carried out based on a multivariate analysis of variance (MANOVA). MANOVA defined the minerals, or volatiles that were significant ($p < 0.05$) for the differentiation of prickly pear juice samples according to geographical origin. The Pillai's trace and Wilks' lambda indices were computed to determine a possible significant effect of minerals, or volatile compounds on the geographical origin of prickly pear fruit juice samples. Linear discriminant analysis (LDA) was then applied only to the selected independent variables by MANOVA, to explore the possibility of differentiating prickly pear juice samples according to geographical origin. For the LDA analysis geographical origin was taken as the dependent variable (grouping variable), while minerals or volatiles were taken as the independent variables [18,20].

More specifically, discriminant analysis creates a predictive model for group membership. The model is composed of one or more discriminant functions based on linear combinations of the predictor variables that provide the best discrimination between the constructed groups. The functions are generated from a sample of cases of known group membership. In addition, the functions can then be applied to the new cases that have measurements for the predictor variables but do not belong in a given group (unknown group membership). In that sense, predicted group membership may be well defined [21]. In addition, regarding the robustness of LDA analysis, Huberty and Olejnik [22] reported studies with a minimum of 10 observations per group. At the same time, what is also important is that the basic criteria (regularity, stability of variations, and independency of variations) should be met. Hence, in the present study these criteria were applied.

Furthermore, a tolerance test was also considered in the analysis. Tolerance is the proportion of a variable's variance not accounted for by other independent variables in the discriminant function developed. A variable with very low tolerance contributes little information to a predictive model and may cause computational problems [23]. Finally, to provide correlations between mineral content of soil samples with those of prickly pear juices samples, Pearson's correlation coefficient (r) was applied at the confidence level $p < 0.05$.

3. Results and Discussion

3.1. Mineral Content Analysis of Soil Samples

During the analysis, significant differences ($p < 0.05$) were observed in total and individual mineral content of soil samples with respect to geographical origin. Full data (average values, mg/L) are given in the supplementary material (Supplementary Table S9). The richest soil in minerals (sum of all minerals, mg/L) was that of the Lakonia region (102,756.15 mg/L), followed by the Eastern (82,246.76 mg/L) and Western Messinia (41,221.04 mg/L) regions, respectively. In addition, the presence of some chemical elements such as Cr, Ti, and Tl in soil samples recorded variations. In particular, Cr was identified in all soil samples, whereas Ti was identified only in soil samples from the region of Lakonia. Finally, Tl was not identified in any of the soil samples analyze. These findings indicate the unique/and or characteristic soil conditions in the soil samples investigated.

3.2. Mineral Content Analysis of Prickly Pear Juice Samples

As in the case of soil samples, the respective mineral content of prickly pear juice samples varied significantly ($p < 0.05$) according to geographical origin. The dominant minerals (mg/kg) were K, P, Ca, Mg, and Na, followed by B, Mn Zn, Sn, and Si. Total mineral content of prickly pear juice samples (average values, mg/kg) was obtained from the sum of each individual mineral and followed the order: Lakonia (2827.98 ± 310.65 mg/kg) > Eastern Messinia (2602.66 ± 203.51 mg/kg) > Western Messinia (2206.07 ± 214.77 mg/kg) (Table 1).

Trace minerals such as Be, Co, Mo, Sb, and Ti were not identified, whereas Cr was identified in two samples (no. 9 and 11 from Lakonia) at 0.03 mg/kg. The same holds for Tl, which was identified in two samples (no. 2) from Western Messinia and Lakonia at 0.04 and 0.20 mg/kg, respectively. Even though considerable amounts of Cr and Ti were identified in the soil samples analyzed (Supplementary Table S9), these elements were identified in minor amounts or were completely absent in the fruit juice samples analyzed. This is probably owed to a specific mineral accumulation mechanism that exists in prickly pear fruit of wild cultivars. On the other hand, despite the fact that Tl was not identified in the soil samples analyzed, this element was identified in prickly pear juice samples (no. 2) from Western Messinia and Lakonia at 0.04 and 0.20 mg/kg, respectively. This "paradox" may be justified by the fact that Tl exists in two oxidation states (+3) and (+1), as ionic salts. The +1 state is more prominent in Tl than the elements above it. Thallium somehow recalls the chemistry of alkali metals. In that sense, thallium (1) ions are found geologically in potassium-based ores, transmitted then to specific fruits, and are probably released through the preparation of prickly pear juice.

Table 1. Mineral content of prickly pear juice of different geographical origin.

Mineral (mg/kg)/Region	Al	B	Ca	Cu	Fe	K	Li	Mg	Mn	Na	Ni	P	Se	Si	Sn	Zn	TMC
Western Messinia (n = 12)																	
Average	0.26 [a]	2.49 [b]	83.78 [c]	0.52 [f]	1.21 [g]	1869.70 [h]	0.12 [k]	93.92 [l]	1.69 [m]	21.36 [o]	0.28 [r]	127.52 [t]	1.23 [w]	0.24 [y]	0.40 [aa]	1.30 [ac]	2206.07 [ad]
±SD	0.26	0.50	9.68	0.35	0.79	205.64	0.10	11.80	2.04	17.41	0.25	17.70	2.05	0.46	0.47	0.56	214.77
Eastern Messinia (n = 12)																	
Average	0.45 [a]	2.24 [b]	59.61 [d]	0.48 [f]	1.36 [g]	2175.97 [i]	0.13 [k]	100.20 [l]	2.83 [n]	48.14 [P]	0.74 [s]	186.13 [u]	0.77 [x]	0.08 [z]	0.14 [ab]	0.79 [ad]	2580.11 [ae]
±SD	1.15	0.45	20.72	0.21	2.33	161.25	0.07	10.18	0.96	25.94	0.21	27.57	0.42	0.18	0.29	0.22	206.06
Lakonia (n = 12)																	
Average	0.27 [a]	2.63 [b]	89.66 [e]	0.67 [f]	0.76 [g]	2398.60 [j]	0.10 [k]	108.50 [l]	1.39 [m]	33.87 [q]	0.35 [r]	188.53 [v]	0.70 [x]	0.24 [y]	0.37 [aa]	1.27 [ac]	2827.98 [af]
± SD	0.38	0.81	33.85	0.39	0.70	252.36	0.10	14.74	2.43	31.10	0.48	28.94	0.38	0.34	0.40	0.33	310.65
LOD	4.90	0.32	4.06	1.40	1.96	1.47	0.12	5.18	0.80	1.45	0.74	1.50	1.49	0.11	8.84	0.40	
LOQ	14.70	1.06	12.17	4.19	5.87	5.41	0.40	15.53	2.40	5.47	2.21	5.00	4.48	0.36	26.53	1.20	
CV (n = 36)	2.15	0.25	0.34	0.59	1.31	0.14	0.78	0.13	0.99	0.79	0.84	0.23	1.35	1.83	1.32	0.40	

N: number of prickly pear juice samples. SD: standard deviation values of three replicates (n = 3). Different letters in each column indicate statistically significant differences ($p <$ 0.05). LOD: limit of detection (µg/kg). LOQ: limit of quantification (µg/kg). CV: coefficient of variation; defined as the ratio of standard deviation to the average, often expressed as a percentage. TMC: total mineral content (mg/kg).

Finally, Ba recorded non-significant differences among the different geographical origins, ranging from 0.03–0.05 mg/kg (average value). Full data regarding the mineral content of minerals determined are given in the supplementary material (Supplementary Table S7). Dehbi et al. [24] reported higher contents of Ca, Mg, P, and Na in prickly pear juice prepared from Moroccan prickly pear cultivars, whereas those of K, Zn, and Cu were lower compared to results of the present study. Higher contents of Ca, K, P, Fe, and Mg were also reported by Mohamed et al. [25] involving Algerian prickly pear juice. However, the respective Na content was much lower compared to present results. The impact of geographical origin is of great importance to realize the differences between previous, even though limited, studies in the literature and results of the present study. Finally, there was also a perfect Pearson's correlation ($r = 1$) ($p < 0.05$) between total mineral content of soil samples with those of prickly pear juice samples.

3.3. Volatile Profile of Prickly Pear Juice

The volatile pattern of prickly pear juice was dominated by alcohols and aldehydes, followed by lower amounts of hydrocarbons and specific terpenoids, such as dl-limonene, along with limited furan derivatives like 2-pentyl-furan. In total, 25 volatile compounds were tentatively identified and then semi-quantified based on the use of the internal standard method (Figure 1). What is remarkable, is that volatile compounds' content (μg/L) of prickly pear juice was affected ($p < 0.05$) by its geographical origin (Table 2). Figure 1a–c represent typical gas chromatograms of prickly pear juice samples from the 3 investigated regions, pointing out the characteristic volatile compounds. Another important issue to note is that the specific alcohols and aldehydes identified in prickly pear juice have been previously reported to contribute to the distinctive aroma of numerous fruits and vegetables [26].

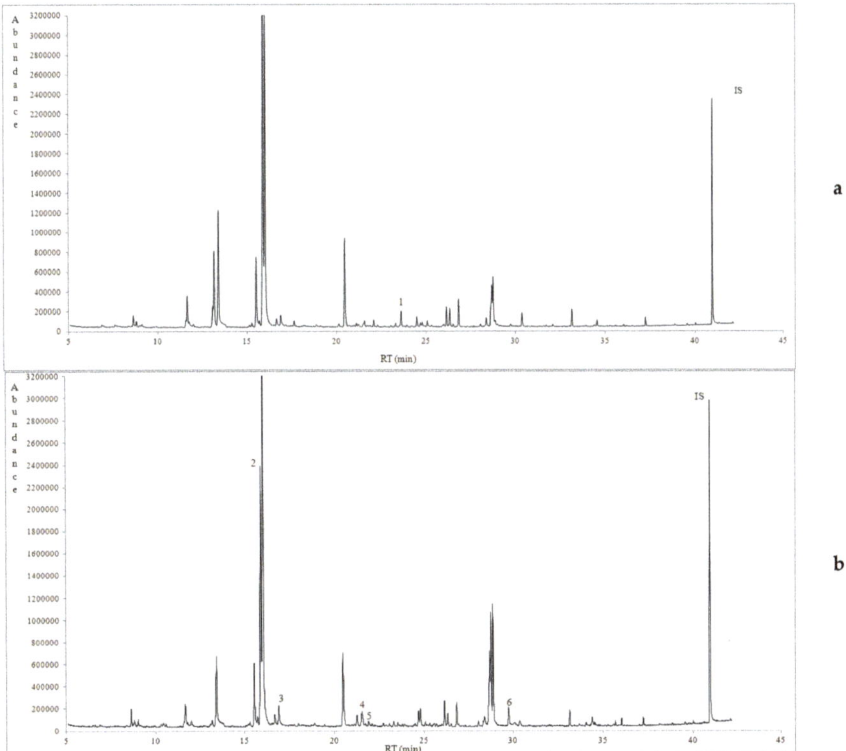

Figure 1. *Cont.*

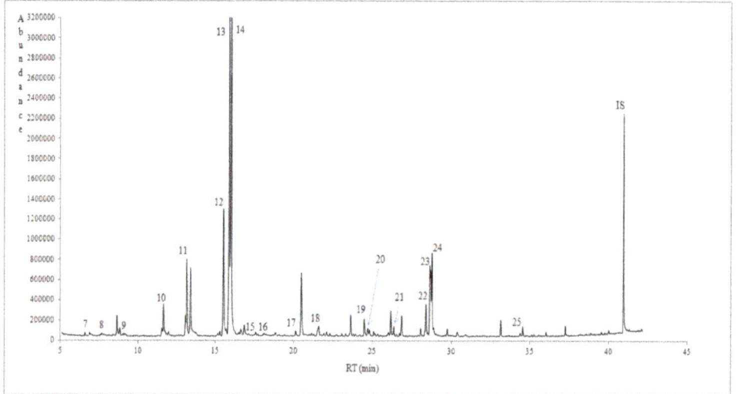

Figure 1. (**a**) A typical gas chromatogram of prickly pear juice sample (no. 12) from Western Messinia. 1: dl-Limonene. IS: internal standard. (**b**) A typical gas chromatogram of prickly pear juice sample (no. 3) from Eastern Messinia. 2: Hexanol. 3: 3,5-Hexadien-1-ol. 4: Decene. 5: Decane. 6: Dodecene. IS: Internal standard. (**c**) A typical gas chromatogram of prickly pear juice sample (no. 11) from Lakonia. 7: 2,4-Hexadiene. 8: 2-Butenal. 9: Pentanal. 10: Pentanol. 11: Hexanal. 12: 3-Hexen-1-ol. 13: 2-Hexenal. 14: 2-Hexen-1-ol. 15: Heptanal. 16: 2,4-Hexadienal. 17: 2-Heptenal. 18: 2-pentyl-Furan. 19: 2-Octenal. 20: Octanol. 21: Nonanal. 22: 2,6-Nonadienal. 23: 2,6-Nonadien-1-ol. 24: 2-Nonen-1-ol. 25: 2,4-Decadienal. IS: Internal standard.

Before going any further, it is important to mention that the formation of flavors in fruits and vegetables may be the outcome of numerous mechanisms, including the lipoxygenase (LOX) pathway and autoxidation reactions [27–29]. In addition, a large number of volatile compounds may be formed in fruits and vegetables during maturation and handling procedures such as cutting, chewing, or during the application of mild heat treatment. In particular, fruits such as apples, pears, peaches, nectarines, apricots, and plums have been reported to have a typical green note when unripe [30,31].

In a previous study dealing with different prickly pear cultivars grown in the region of Paterno (Catania, Italy), Arena et al. [32] reported that the volatile profile recorded using HS-SPME/GC-MS was dominated by different amounts (μg/kg) of (E)-2-hexenal, (Z)-2-penten-1-ol, hexan-1-ol, (Z)-3-hexen-1-ol, (E)-2-hexen-1-ol, (E)-2-nonen-1-ol and (E,Z)-2,6-nonadien-1-ol, and trace amounts of (E)-2-nonenal and (E,Z)-2,6-nonadienal identified in a few samples. These findings are in agreement with the results of the present study.

Oumato et al. [33] using HS-SPME/GC-MS, reported that the dominant volatile compounds identified in three different prickly pear cultivars grown in the wider area of Morocco were 2-hexanal and 1-hexanol, among numerous others, also identified in the present study.

The compound 2,6-Nonadienal has attracted great attention as the essence of cucumbers [34] but it was also found in freshly cut watermelon [35]. Compounds such as (E)-2-hexen-1-ol and (E,E)-2,4-decadienal have been reported to dominate the aroma of peaches (*P. persica*), nectarines (*P.persica* var. nucipersica), and sweet cherries (*P. avium*) [36,37]. On the other hand, 1-hexanol and 2-octenal contributed to the aroma of peas (*Pisum sativum*) [38]. A similar study on Egyptian prickly pear juice, or its blends with mandarin juice, highlighted that hexanal, nonanal, octanal, and (E)-2-hexenal (among other volatiles) contributed to its aroma [25]. This is in agreement with present results.

Another characteristic volatile compound, dl-Limonene, which represents the D-isomer of limonene, is a natural occurring volatile that is responsible for the characteristic flavor of citrus fruits. Limonene is formed from geranyl pyrophosphate, via cyclization of a neryl carbocation or its equivalent. The final step involves loss of a proton from the cation to form the alkene [39].

Table 2. Volatile compounds of prickly pear juice tentatively identified and semi-quantified according to geographical origin.

RT (min)	Volatile Compounds (µg/L)	KI [a]	Western Messinia (Avg ± SD)	Eastern Messinia (Avg ± SD)	Lakonia (Avg ± SD)	Method of Identification	MS Qualification (%)	Orthonasal Threshold (µg/L) [b]	Odour Note [c]
	Alcohols								
11.66	1-Pentanol	702	192.54 ± 90.69	178.18 ± 54.94	314.94 ± 105.62	MS	90	0.0055-305	Fermented, green
15.27	3-Hexen-1-ol	787	32.26 ± 15.86	41.02 ± 31.12	47.81 ± 21.01	MS	94	na	Woody, green, leafy
15.89	2-Hexen-1-ol	802	2486.91 ± 593.02	2681.84 ± 1280.63	3022.18 ± 1101.85	MS/KI	91	na	Green, fruity, leafy
16.02	1-Hexanol	804	2372.75 ± 1341.20	3885.41 ± 941.84	3429.98 ± 1149.45	MS/KI	90	500-2500	Green, fruity
16.86	3,5-Hexadien-1-ol	823	120.61 ± 57.82	189.66 ± 53.18	180.42 ± 67.70	MS/KI	92	na	
24.79	1-Octanol	999	ni	ni	55.27 ± 46.57	MS/KI	90	190	Herbal, green, penetrating
28.67	2,6 Nonadien-1-ol	1090	318.22 ± 265.20	270.39 ± 222.85	797.17 ± 337.37	MS/KI	91	na	Green, cucumber-like
28.75	2-Nonen-1-ol	1091	443.17 ± 240.80	564.97 ± 124.63	921.67 ± 240.39	MS/KI	92	na	Fatty
	Aldehydes								
7.63	2-Butenal	520	ni	ni	41.58 ± 19.44	MS	90	na	Floral
9.13	Pentanal	620	ni	ni	19.87 ± 18.71	MS	86	na	Bready, fruity, berry-like
13.17	Hexanal	737	431.69 ± 218.41	288.86 ± 241.71	692.64 ± 156.50	MS	96	9.18-10.50	Green, grassy, floral
15.51	2-Hexenal	793	512.87 ± 263.34	627.65 ± 411.57	1129.38 ± 258.88	MS	96	24.2	Soapy, fatty, green
17.63	Heptanal	840	26.20 ± 17.26	ni	43.79 ± 46.45	MS/KI	90	na	Fruity, oily-greasy
18.14	2,4-Hexadienal	851	ni	ni	27.72 ± 23.07	MS/KI	92	na	Green, fruity, waxy
20.11	2-Heptenal	894	29.28 ± 28.21	ni	55.93 ± 22.71	MS/KI	97	na	Green, fatty, oily, fruity
24.48	2-Octenal	992	64.02 ± 49.55	ni	181.08 ± 48.33	MS/KI	91	na	Sweet, green, fatty, brothy
26.34	Nonanal	1035	86.00 ± 35.76	96.35 ± 20.97	187.25 ± 58.03	MS/KI	91	2.53-5.00	Soapy, floral
28.37	2,6-Nonadienal	1082	78.21 ± 70.68	80.75 ± 68.90	378.80 ± 152.75	MS/KI	90	na	Cucumber-like, green
34.37	2,4-Decadienal	1243	ni	ni	54.09 ± 47.75	MS/KI	93	0.2	Fatty, waxy, green
	Hydrocarbons								
6.95	2,4-Hexadiene	473	ni	ni	12.66 ± 13.81	MS	90	na	na
21.52	1-Decene	926	ni	75.67 ± 65.39	ni	MS/KI	95	6.45	Pleasant
21.89	Decane	934	ni	22.57 ± 20.78	ni	MS/KI	93	na	na
29.73	1-Dodecene	1116	ni	64.01 ± 60.12	ni	MS/KI	95	na	na
	Terpenoids								
23.56	dl-Limonene	971	23.07 ± 49.07	14.85 ± 9.83	ni	MS/KI	98	0.0018-0.31	Lemon, citrus
	Furan derivatives								
21.58	Furan, 2-pentyl-	927	40.71 ± 35.39	ni	109.03 ± 81.45	MS/KI	91	10.06	Strong

RT: retention time (min). [a] KI: Kovats index. Avg ±SD: average value ±standard deviation. MS: mass spectra. ni: not identified. na: not available. [b]: Berger [26], AIHA [40]. [c]: Berger [26], Burdock [41].

Table 2 also contains the orthonasal threshold value (OTV) for some volatile compounds available in the literature. OTV may be defined as the minimum amount of a compound that can be detected by the human nose. Aroma compounds are volatile compounds which are strongly perceived by the odor receptor sites of the olfactory tissue of the nasal cavity. These compounds directly reach the odor receptors when they "pass" through the nasal cavity (orthonasal detection) [42]. Based on available OTV literature data (Table 2), present results indicate that prickly pear juice is dominated by specific and undisputed aroma compounds. At this point, it should be mentioned that there is no study in the literature reporting data on the volatile profile of prickly pear juice prepared from prickly pear fruits grown in different areas of Peloponnese/Greece. Considering the above, the aroma of prickly pear juice is most probably the synergistic outcome of numerous volatiles.

3.4. Valorization of Prickly Pear Juice Geographical Origin Based on Mineral Content Using LDA

MANOVA analysis identified the significant minerals that could be used for the geographical discrimination of prickly pear juice samples. The sixteen minerals were considered as the dependent variables, while the three regional zones (geographical origin) were taken as the independent variables. The two qualitative criteria of multivariate statistics namely Pillai's trace = 1.694 (F = 6.571, df = 32, $p < 0.001$) and Wilks' lambda = 0.022 (F = 6.469, df = 32, $p < 0.001$), showed that there was a statistically significant effect of prickly pear juice mineral content on the geographical origin of fruit juices. In particular, 7 of the 16 minerals (Table 3) were found to be significant ($p < 0.05$) for the geographical discrimination of prickly pear juices. These minerals were, then, subjected to LDA. In addition, the minimum tolerance level of the analysis was set at 0.001. Results showed that Fe did not pass the tolerance test. Therefore, it was excluded (SPSS program) a priori from the discriminant analysis. During LDA analysis two canonical discriminant functions were formed: Wilks' lambda = 0.063, $X^2 = 80.320$, df = 14, $p < 0.001$ for the first function and Wilks' lambda = 0.348, $X^2 = 30.611$, df = 6, $p < 0.001$, for the second. The first discriminant function recorded the higher eigenvalue (4.552) and canonical correlation of 0.905, accounting for 70.8% of total variance. The second discriminant function recorded a lower eigenvalue (1.874) and canonical correlation of 0.807, accounting for 29.2% of total variance. Both accounted for 100% of total variance.

Table 3. Contribution of minerals to the canonical discriminant function structure matrix.

Minerals	Wilks' Lambda	F	df1	df2	p	Function 1	Function 2
Ca	0.748	5.397	2	32	0.0010	0.132	0.371 *
K	0.458	18.938	2	32	<0.001	0.463 *	−0.334
Mg	0.797	4.070	2	32	0.027	0.230 *	−0.086
Na	0.770	4.768	2	32	0.015	0.025	−0.397 *
Ni	0.746	5.439	2	32	0.009	−0.047	−0.419 *
P	0.428	21.397	2	32	<0.001	0.381	−0.600 *
Zn	0.704	6.741	2	32	0.004	0.081	0.457 *

F: function value, df: degrees of freedom, p: level of significance. Pooled within-groups correlations between discriminating variables and standardized canonical discriminant functions. Variables ordered by absolute size of correlation within function. * Largest absolute correlation between each variable and any discriminant function.

Prickly pear fruit juices are well separated, as shown in Figure 2a. The classification rate was 94.3% using the original, and 85.7% using the cross-validation method. The geographical classification rate was 91.7%, 81.8%, and 83.3% for Western Messinia, Eastern Messinia, and Lakonia, respectively (Table S10). The group centroid values, characteristic for each geographical origin, are also pointed out in Figure 2a. What should be defined, is that each centroid has two numbers which represent the coordinates. The abscissa is the first discriminant function and the ordinate is the second. The respective group centroid values were: (−1.899, 1.341), (−0.954, −1.834), and (2.774, 0.340) for the Western Messinia, Eastern Messinia, and Lakonia regions. The farther apart the means are, the less error there will be in classification [17].

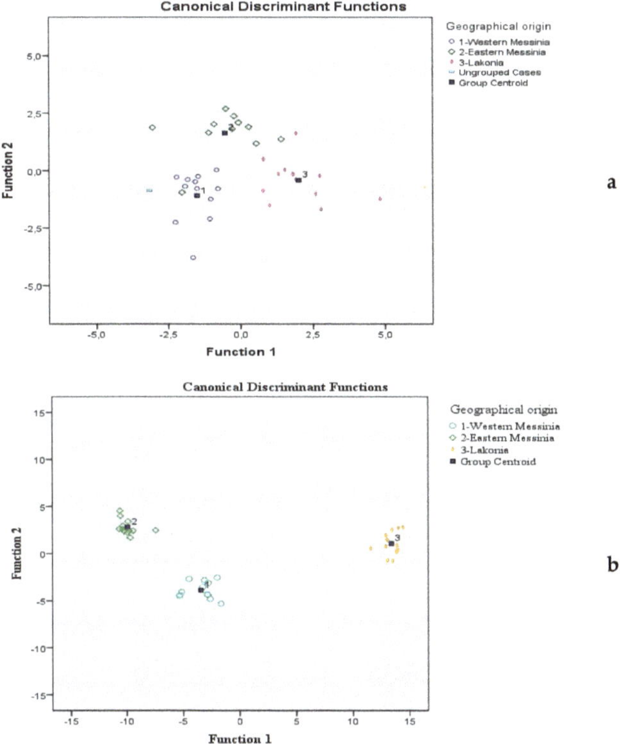

Figure 2. (**a**) Classification of prickly pear juice according to geographical origin based on 7 minerals and linear discriminant analysis (LDA). (**b**) Classification of prickly pear juice according to geographical origin based on 21 volatile compounds and LDA.

Simpkins et al. [9] used inductively coupled plasma atomic emission spectroscopy (ICP-AES) and inductively coupled plasma mass spectrometry for the determination of minerals and trace elements (Al, B, Ba, Ca, Co, Cu, Fe, K, Li, Lu, Mg, Mn, Mo, Na, Ni, P, Rb, Si, Sr, Sn, Ti, V, and Zn) in Australian and Brazilian orange juice samples. Principal component analysis (PCA) in mineral and trace element contents resulted in a clear differentiation of orange juice samples. In another study, Pellerano et al. [13] used instrumental neutronic activation analysis (INAA) for the determination of Br, As, Na, Rb, La, Cr, Sc, Fe, Co, Zn, and Sb in lemon juice samples collected from three different geographical origins in the northwest region of Argentina. Application of PCA and LDA in the mineral content of lemon juice samples resulted in their correct classification according to geographical origin by 93.2%; higher by 7.5% compared to present results.

3.5. Valorization of Prickly Pear Juice Geographical Origin Based on Semi-Quantitative Data of Volatile Compounds Using LDA

Similarly, Pillai's trace = 1.932 (F = 11.444, df = 50, $p < 0.001$) and Wilks' lambda = 0.000 (F = 16.335, df = 50, $p < 0.001$) index values showed that there was a significant multivariable effect of volatile compounds on geographical origin of prickly pear juice. Twenty-two volatile compounds were found to be significant ($p < 0.05$) for the geographical discrimination of prickly pear juice samples. To avoid any misleading source, it should be stressed that 2-pentyl-furan, despite the significant differences ($p < 0.05$) among prickly pear juice samples according to geographical origin, was not included in the statistical analysis because it was categorized as a thermal artifact. Therefore, the semi-quantitative data of 21 volatile compounds was subjected to LDA.

Two canonical significant discriminant functions, as in the case of minerals, were formed: Wilks' Lambda = 0.001, X^2 = 159.643, df = 42, p < 0.001 for the first and Wilks' Lambda = 0.103, X^2 = 52.338, df = 20, p < 0.001 for the second function, respectively. The first discriminant function recorded the higher eigenvalue (105.215) and canonical correlation of 0.995, accounting for 92.3% of total variance. The second discriminant function recorded a much lower eigenvalue (8.733) and canonical correlation of 0.947, accounting for 7.7% of total variance. Both accounted for 100% of total variance which is a perfect rate. As it may be seen in Figure 2b, prickly pear fruit juices are clearly separated. The correct classification rate was 100% using the original and 88.9% using the cross-validation method. The geographical classification rate was 75%, 91.7%, and 100% for Western Messinia, Eastern Messinia and Lakonia, respectively (Table S11). Respective group centroid values were: (−3.469, −3.875), (−9.912, 2.803), and (13.381, 1.072) for Western Messinia, Eastern Messinia and Lakonia regions. The volatile compounds which served as markers of prickly pear juice geographical origin are listed in Table 4. However, differences in the volatile profile and as a consequence, in the classification rates obtained, may also be attributed to genetic factors (i.e., cultivar) [32,33], given the fact that wild-grown prickly pear cultivars from specific regions were used in the study.

Table 4. Contribution of volatile compounds to the canonical discriminant function structure matrix.

Volatile Compounds (µg/L)	Wilks' Lambda	F	df1	df2	p	Function 1	Function 2
2-Butenal	0.231	54.909	2	33	<0.001	0.171 **	0.165
2,4-Decadienal	0.517	15.401	2	33	<0.001	0.091 **	0.088
Decanal	0.538	14.158	2	33	<0.001	−0.064	0.220 **
Decene	0.507	16.070	2	33	<0.001	−0.069	0.234 *
1-Dodecene	0.548	13.604	2	33	<0.001	−0.063	0.215 *
Heptanal	0.699	7.120	2	33	0.003	0.060	−0.079 **
2-Heptenal	0.434	21.488	2	33	<0.001	0.107 **	−0.107
2,4-Hexadienal	0.488	17.326	2	33	<0.001	0.096 **	0.093
2,4-Hexadiene	0.620	10.094	2	33	<0.001	0.073 **	0.071
3,5-Hexadien-1-ol	0.778	4.703	2	33	0.016	0.007	0.179 **
Hexanal	0.588	11.554	2	33	<0.001	0.081 **	−0.024
1-Hexanol	0.753	5.410	2	33	0.009	−0.003	0.194 **
2-Hexenal	0.566	12.657	2	33	<0.001	0.077	0.128 **
2,6-Nonadienal	0.337	32.509	2	33	<0.001	0.132 **	0.131
2,6-Nonadien-1-ol	0.558	13.067	2	33	<0.001	0.085 **	0.057
Nonanal	0.429	21.969	2	33	<0.001	0.105	0.139 **
2-Nonen-1-ol	0.493	16.955	2	33	<0.001	0.086	0.170 **
1-Octanol	0.494	16.903	2	33	<0.001	0.095 **	0.092
2-Octenal	0.207	63.348	2	33	<0.001	0.190 **	−0.056
Pentanal	0.549	13.541	2	33	<0.001	0.085 **	0.082
Pentanol	0.645	9.079	2	33	0.001	0.071 **	0.044

F: function value, df: degrees of freedom, p: level of significance. Pooled within-groups correlations between discriminating variables and standardized canonical discriminant functions. Variables ordered by absolute size of correlation within function. ** Largest absolute correlation between each variable and any discriminant function.

In a relevant study, Reid et al. [10] used SPME-GC/MS for the determination of volatile compounds in apple juice prepared from different varieties. Further application of LDA in the volatile compounds data, resulted in the correct classification rate of apple juice samples by 87.5%. More recently, application of SPME-GC/MS and principal component analysis on the volatile compounds identified in tomato juice from Italian and Spanish markets resulted in the explanation of 68.61% of total variance among the samples analyzed [43].

3.6. Summary Regarding the Most Effective Predictors of Prickly Pear Juice Geographical Origin

In Tables 3 and 4 are listed with an asterisk (among other statistical analysis parameters) the standardized canonical discriminant function coefficients obtained in the selected models for every mineral and volatile compound according to prickly pear juice geographical origin. What is of great interest is that the higher the absolute value of a standardized canonical coefficient, the more significant the variable [17]. In both cases (analyses of minerals and volatile compounds) the first discriminant function was the one that differentiated best between prickly pear juice groups, given that it represented the highest variability (70.8% and 92.3%, respectively). However, the contribution of the second discriminant function is of considerable value, since both discriminant functions explained 100% of total variance. Based on the aforementioned, specific minerals and volatile compounds may effectively assist in the valorization of prickly pear juice from the Peloponnese Peninsula.

4. Conclusions

Prickly pear juice prepared from wild cultivars grown in the region of Peloponnese is a good source of micro- and macro-minerals and may serve as a beneficial fruit beverage in the diet. On the other hand, prickly pear juice proved to have a balanced and distinctive aroma dominated by alcohols and aldehydes. A total of 7 minerals and 21 volatile compounds provided satisfactory classification rates of prickly pear juice samples according to geographical origin when subjected to LDA, and are proposed as "geographical origin indicators" of prickly pear juice from the Peloponnese Peninsula. The exploitation of prickly pear juice may support the local economy and contribute to the preparation of added value products of specific origin.

Supplementary Materials: The following are available online at http://www.mdpi.com/2304-8158/8/4/123/s1, Table S1: Linearity of the calibration curves used for the quantification of minerals in soil samples, Table S2: Linearity of the calibration curves used for the quantification of minerals in prickly pear juice samples, Table S3a: Mineral recoveries in soil samples at 2 mg/kg, Table S3b: Mineral recoveries in soil samples at 200 mg/kg, Table S4: Mineral recoveries in prickly pear juice samples at different spiking concentrations, Table S5: Limit of detection (LOD) and limit of quantification (LOQ)* in soil samples, Table S6: Limit of detection (LOD) and limit of quantification (LOQ) (µg/kg) in fruit juice samples, Table S7: Mineral content (mg/kg) of prickly pear juice samples according to geographical origin, Table S8: SPME-GC/MS method optimization/development, Table S9: Mineral content (mg/L) of soil samples according to prickly pear geographical origin, Table S10: Discriminatory ability of the developed LDA model for the classification of prickly pear juice according to geographical origin based on 7 minerals, Table S11: Discriminatory ability of the developed LDA model for the classification of prickly pear juice according to geographical origin based on 21 volatile compounds, Figure S1: A typical gas chromatogram of a prickly pear juice mixture of the 3 regions during method optimization. TEST1, TEST2, and TEST 4 refer to methods 1, 2, 3 (Table S8), Figure S2: A typical gas chromatogram of a prickly pear juice mixture of the 3 regions during method optimization. TEST5, TEST6, and TEST7 refer to method 3 (Table S8), Figure S3: A typical gas chromatogram of a prickly pear juice mixture of the 3 regions during method optimization. TEST8, TEST9, and TEST10 refer to method 3 (Table S8), Figure S4: A typical gas chromatogram of a prickly pear juice mixture of the 3 regions during method optimization. TEST11, TEST12, and TEST13 refer to method 3 (Table S8), Figure S5: A typical gas chromatogram of a prickly pear juice mixture of the 3 regions during method optimization. TEST14, TEST15, and TEST17 refer to method 3 (Table S8), Figure S6: A typical gas chromatogram of a prickly pear juice mixture of the 3 regions during method optimization. TEST7, TEST10, and TEST13 refer to method 3 (Table S8), Figure S7: A typical gas chromatogram of a prickly pear juice mixture of the 3 regions during method optimization. TEST13 and TEST17 refer to method 3 (Table S8), Figure S8: A typical gas chromatogram of a prickly pear juice mixture of the 3 regions during method optimization. TEST20 and TEST22 refer to method 3 (Table S8).

Author Contributions: Conceptualization. I.K.K.; Methodology, V.K.K., I.K.K., A.L.; Software, A.B., C.P.; Validation, V.K.K., I.K.K. and A.L.; Formal analysis, V.K.K., I.K.K., A.L.; Investigation, V.K.K., I.K.K.; Resources, A.B., M.G.K., C.P.; Data interpretation, V.K.K., I.K.K.; Writing—original draft preparation, I.K.K.; Writing—review and editing, I.K.K., M.G.K; Visualization, I.K.K., V.K.K.; Supervision, I.K.K.; Project administration, I.K.K.

Funding: This research received no external funding.

Conflicts of Interest: The authors declare no conflicts of interest.

References

1.   Quattrocchi, U. *CRC World Dictionary of Plant Names*; CRC Press, Taylor & Francis Group: Boca Raton, FL, USA, 2000; Volume III, p. 1885.

2. United States Department of Agriculture (USDA). Food Description: "09287, Prickly Pears, Raw". 2017. Available online: https://www.usda.gov (accessed on 24 September 2018).

3. Butera, D.; Tesoriere, L.; Di Gaudio, F.; Bongiorno, A.; Allegra, M.; Pintaudi, A.M.; Kohen, R.; Livrea, A.M. Antioxidant activities of sicilian prickly pear (*Opuntia ficus indica*) fruit extracts and reducing properties of its betalains: Betanin and indicaxanthin. *J. Agric. Food Chem.* **2002**, 50, 6895–6901. [CrossRef]

4. Guzmán-Maldonado, S.H.; Morales-Montelongo, A.L.; Mondragón-Jacobo, C.; Herrera-Hernández, G.; Guevara-Lara, F.; Reynoso-Camacho, R. Physicochemical, nutritional, and functional characterization of fruits xoconostle (*Opuntia matudae*) pears from Central-México Region. *J. Food Sci.* **2010**, 75, C485–C492. [CrossRef] [PubMed]

5. Frati, A.C.; Xilotl Díaz, N.; Altamirano, P.; Ariza, R.; López-Ledesma, R. The effect of two sequential doses of *Opuntia streptacantha* upon glycemia. *Archivos De Investigación Médica* **1991**, 22, 333–336. [PubMed]

6. Franke, A.A.; Cooney, R.V.; Henning, S.M.; Custer, L.J. Bioavailability and antioxidant effects of orange juice components in humans. *J. Agric. Food Chem.* **2005**, 53, 5170–5178. [CrossRef]

7. Zielinski, A.A.F.; Haminium, C.W.I.; Nunes, C.A.; Schnitzler, E.; van Ruth, S.M.; Granato, D. Chemical composition, sensory properties, provenance, and bioactivity of fruit juices as assessed by chemometrics: A critical review and guideline. *Compr. Rev. Food Sci. Food Saf.* **2014**, 13, 300–316. [CrossRef]

8. Mangas, J.J.; Suarez, B.; Picinelli, A.; Moreno, J.; Blanco, D. Differentiation by phenolic profile of apple juices prepared according to membrane techniques. *J. Agric. Food Chem.* **1997**, 45, 4777–4784. [CrossRef]

9. Simpkins, W.A.; Louie, H.; Wu, M.; Harrison, M.; Goldberg, D. Trace elements in Australian orange juices and others products. *Food Chem.* **2000**, 71, 423–433. [CrossRef]

10. Reid, L.M.; O'Donnell, C.P.; Kelly, J.D.; Downey, G. Preliminary studies for the differentiation of apple juice samples by chemometric analysis of solid-phase microextraction-gas chromatographic data. *J. Agric. Food Chem.* **2004**, 52, 6891–6896. [CrossRef]

11. He, J.; Rodrigues-Saona, L.E.; Giusti, M. Mid-infrared spectroscopy for juice authentication-rapid differentiation of commercial juices. *J. Agric. Food Chem.* **2007**, 55, 4443–4452. [CrossRef]

12. Martina, V.; Ionescu, K.; Pigani, L.; Terzi, F.; Ulrici, A.; Zanardi, C.; Seeber, R. Development of an electronic tongue based on a PEDOT-modified voltammetric sensor. *Anal. Bioanal. Chem.* **2007**, 387, 2101–2110. [CrossRef]

13. Pellerano, R.G.; Mazza, S.S.; Marigliano, R.A.; Marchevsky, E.J. Multielement analysis of Argentinean lemon juices by instrumental neutronic activation analysis and their classification according to geographical origin. *J. Agric. Food Chem.* **2008**, 56, 5222–5225. [CrossRef]

14. Longobardi, F.; Ventrella, A.; Bianco, A.; Catucci, L.; Cafagna, I.; Gallo, V.; Mastrorilli, P.; Agostiano, A. Non-targeted ^{1}H NMR fingerprinting and multivariate statistical analyses for the characterization of the geographical origin of Italian sweet cherries. *Food Chem.* **2013**, 141, 3028–3033. [CrossRef]

15. Karabagias, I.K.; Louppis, P.A.; Kontakos, S.; Papastephanou, C.; Kontominas, M.G. Characterization and geographical discrimination of Greek pine and thyme honeys based on their mineral content, using chemometrics. *Eur. Food Res. Technol.* **2017**, 243, 101–113. [CrossRef]

16. AOAC Official Method 990.08. *Metals in Solid Wastes, Inductively Coupled Plasma Atomic Emission Spectrometric Method*; AOAC International: Rockville, MD, USA, 1993.

17. Karabagias, I.K.; Louppis, P.A.; Karabournioti, S.; Kontakos, S.; Papastephanou, C.; Kontominas, M.G. Characterization and geographical discrimination of commercial Citrus spp. honeys produced in different Mediterranean countries based on minerals, volatile compounds and physicochemical parameters, using chemometrics. *Food Chem.* **2017**, 217, 445–455. [CrossRef]

18. Karabagias, I.K.; Badeka, A.; Kontakos, S.; Karabournioti, S.; Kontominas, M.G. Characterisation and classification of Greek pine honeys according to their geographical origin based on volatiles, physicochemical parameters and chemometrics. *Food Chem.* **2014**, 146, 548–557. [CrossRef] [PubMed]

19. Karabagias, I.K. Volatile metabolites or pollen characteristics as regional markers of monofloral thyme honey? *Sep. Sci. Plus* **2018**, 1, 83–92. [CrossRef]

20. Louppis, P.A.; Karabagias, I.K.; Kontakos, S.; Kontominas, M.G.; Papastephanou, C. Botanical discrimination of Greek unifloral honeys based on mineral content in combination with physicochemical parameter analysis, using a validated chemometric approach. *Microchem. J.* **2017**, 135, 180–189. [CrossRef]

21. Miller, J.N.; Miller, J.C. *Statistics and Chemometrics for Analytical Chemistry*, 6th ed.; Pearson Education Limited: Essex, UK, 2010; pp. 1–278.

22. Huberty, C.J.; Olejnik, S. *Applied MANOVA and Discriminant Analysis*, 2nd ed.; John Wiley & Sons: Hoboken, NJ, USA, 2006; pp. 355–356.

23. Krishnamoorthy, K. *Statistical Tolerance Regions: Theory, Applications, and Computation*; John Wiley & Sons: Hoboken, NJ, USA, 2009; pp. 1–6.

24. Dehbi, F.; Hasib, A.; Ouatmane, A.; Elbatal, H.; Jaouad, A. Physicochemical characteristics of Moroccan prickly pear juice (*Opuntia ficus indica* L.). *Int. J. Adv. Res. Technol.* **2014**, *4*, 300–306.

25. Mohamed, S.A.; Hussein, A.M.S.; Ibraheim, G.E. Physicochemical, sensorial, antioxidant and volatile of juice from prickly pear with guava or mandarin. *Int. J. Food Sci. Nutr.* **2014**, *3*, 44–53.

26. Berger, R.F. *Flavours and Fragrances, Chemistry, Bioprocessing and Sustainability*; Springer: Berlin/Heidelberg, Germany, 2007; pp. 1–649.

27. Yilmaz, E.; Baldwin, E.A.; Shewfelt, R.L. Enzymatic modification of tomato Homogenate and its effect on volatile flavor Compounds. *J. Food Sci.* **2002**, *67*, 2122–2125. [CrossRef]

28. Haslbeck, F.; Grosch, W. HPLC analysis of all positional isomers of the monohydroperoxides formed by soybean lipoxygenases during oxidation of linoleic acid. *J. Food Biochem.* **2007**, *9*, 1–14. [CrossRef]

29. Chan, H.W.S. *Autoxidation of Unsaturated Lipids*; Academic Press: London, UK, 1987; pp. 1–296.

30. Engel, K.H.; Ramming, D.W.; Flath, R.A.; Teranishi, R. Investigation of volatile constituents in nectarines. 2. Changes in aroma composition during nectarine maturation. *J. Agric. Food Chem.* **1988**, *36*, 1003–1006. [CrossRef]

31. Aubert, C.; Günata, Z.; Ambid, C.; Baumes, R. Changes in physicochemical characteristics and volatile constituents of yellow- and white-fleshed nectarines during maturation and artificial ripening. *J. Agric. Food Chem.* **2003**, *51*, 3083–3091. [CrossRef]

32. Arena, E.; Campisi, S.; Fallico, B.; Lanza, M.C.; Maccarone, E. Aroma value of volatile compounds of prickly pear (*Opuntia ficus indica* L. Mill. Cactaceae). *Int. J. Food Sci. Nutr.* **2001**, *3*, 311–319.

33. Oumato, J.; Zrira, S.; Petretto, G.L.; Saidi, B.; Salaris, M.; Pintore, G. Volatile constituents and polyphenol composition of *Opuntia ficus indica* (L.) Mill from Morocco. *Dir. Open Access J.* **2016**, *4*, 5–11.

34. Schieberle, P.; Ofner, S.; Grosch, W. Evaluation of potent odorants in cucumbers (*Cucumis sativus*) and Muskmelons (*Cucumis melo*) by aroma extract dilution analysis. *J. Food Sci.* **1990**, *55*, 193–195. [CrossRef]

35. Guler, Z.; Candir, E.; Yetisir, H.; Karaca, F.; Solmaz, I. Volatile organic compounds in watermelon (Citrullus lanatus) grafted onto 21 local and two commercial bottle gourd (*Lagenaria siceraria*) rootstocks. *J. Hortic. Sci. Biotechnol.* **2014**, *89*, 448–452. [CrossRef]

36. Maarse, H. *Volatile Compounds in Fruits and Beverages*; Dekker: New York, NY, USA, 1991.

37. Lavilla, T.; Recasens, I.; Lopez, M.L.; Puy, J. Multivariate analysis of maturity stages, including quality and aroma, in 'Royal Glory' peaches and 'Big Top' nectarines. *J. Sci. Food Agric.* **2002**, *82*, 1842–1849. [CrossRef]

38. Jakobsen, H.B.; Hansen, M.; Chrisyensen, M.R.; Brockhoff, P.B.; Olsen, C.E. Aroma volatiles of blanched green peas (*Pisum sativum* L.). *J. Agric. Food Chem.* **1998**, *46*, 3727–3734. [CrossRef]

39. Mann, J.C.; Hobbs, J.B.; Banthorpe, D.V.; Harborne, J.B. *Natural Products: Their Chemistry and Biological Significance*; Longman Scientific & Technical: Essex, UK, 1994; pp. 308–309.

40. American Industrial Hygiene Association (AIHA). *Odor Thresholds for Chemicals with Established Health Standards*, 2nd ed.; AIHA: Fairfax, VA, USA, 2013.

41. Burdock, G.A. *Fenaroli's Handbook of Flavor Ingredients*, 5th ed.; CRC Press: Boca Raton, FL, USA, 2004; pp. 1–1864.

42. Belitz, H.D.; Grosch, W.; Schieberle, P. *Food Chemistry*, 4th ed.; Springer: Berlin/Heidelberg, Germany, 2009; pp. 1–988.

43. Vallverdú-Queralt, A.; Bendini, A.; Tesini, F.; Valli, E.; Lamuela-Raventos, R.M.; Toschi, T.G. Chemical and sensory analysis of commercial tomato juices present on the Italian and Spanish markets. *J. Agric. Food Chem.* **2013**, *61*, 1044–1050. [CrossRef] [PubMed]

Article

Targeted and Untargeted Metabolomics as an Enhanced Tool for the Detection of Pomegranate Juice Adulteration

Marilena E. Dasenaki *, Sofia K. Drakopoulou, Reza Aalizadeh and Nikolaos S. Thomaidis

Laboratory of Analytical Chemistry, Department of Chemistry, National and Kapodistrian University of Athens, Panepistimiopolis Zographou, 15771 Athens, Greece; sofiadrakop@chem.uoa.gr (S.K.D.); raalizadeh@chem.uoa.gr (R.A.); ntho@chem.uoa.gr (N.S.T.)
* Correspondence: mdasenaki@chem.uoa.gr; Tel.: +30-210-7274430; Fax: +30-210-7274750

Received: 29 May 2019; Accepted: 12 June 2019; Published: 14 June 2019

Abstract: Pomegranate juice is one of the most popular fruit juices, is well-known as a "superfood", and plays an important role in healthy diets. Due to its constantly growing demand and high value, pomegranate juice is often targeted for adulteration, especially with cheaper substitutes such as apple and red grape juice. In the present study, the potential of applying a metabolomics approach to trace pomegranate juice adulteration was investigated. A novel methodology based on high-resolution mass spectrometric analysis was developed using targeted and untargeted screening strategies to discover potential biomarkers for the reliable detection of pomegranate juice adulteration from apple and red grape juice. Robust classification and prediction models were built with the use of unsupervised and supervised techniques (principal component analysis (PCA) and partial least squares discriminant analysis (PLS-DA)), which were able to distinguish pomegranate juice adulteration to a level down to 1%. Characteristic *m/z* markers were detected, indicating pomegranate juice adulteration, and several marker compounds were identified. The results obtained from this study clearly demonstrate that Mass Spectrometry (MS)-based metabolomics have the potential to be used as a reliable screening tool for the rapid determination of food adulteration.

Keywords: fruit juice authenticity; pomegranate juice; adulteration; high-resolution mass spectrometry; biomarkers

---

1. Introduction

The fruit and vegetable juice industry is one of the world's fastest-growing segments of the beverage industry due to the mounting focus of consumers on a healthy and balanced diet. As health and fitness have become vital in today's world, changes in lifestyles have steered the growth of the global juice market across various developing and developed countries. According to the "Global Fruit and Vegetable Juice Market Research Report, 2018–2025", the value of the global fruit juice market reached 154.18 billion USD in 2016 and is expected to grow at a compound annual growth rate (CAGR) of 5.93% during the forthcoming years [1].

Compared to other types of fruit juice, the popularity of pomegranate juice has skyrocketed in the last decade, mainly due to its well-established health benefits [2]. The pomegranate (*Punica granatum* L.) is an excellent source of precious nutrients, such as vitamins, sugars, acids, polysaccharides, polyphenols, and minerals, promoting an organism's health and wellness. On top of that, pomegranate juice's antioxidant activity has been repeatedly reported to be higher compared to that of other fruit juices [3]. Regular pomegranate juice consumption has been linked with the improvement of cardiovascular health through the reduction of cholesterol levels and the lowering of blood pressure, and also with the prevention of skin, breast, and prostate cancer [3–5]. With antimicrobial, anti-inflammatory, astringent, antitussive, and antidiarrheal properties, pomegranate juice has gained a reputation as an easily accessible superfood and is being sold as a high-quality food item [2].

Pomegranate juice's economic value, along with its constantly increasing demand, which often exceeds supply, makes it vulnerable to adulteration [3]. The adulteration of pomegranate juice mostly includes dilution with water, the addition of sugars, or mixing with cheaper juices (apple, grape, pear) and is an illegal practice adopted by several suppliers and manufacturers to compensate for high product demand and mask low-quality raw materials [6]. This practice has a negative impact on the nutritional value of the juice and may also involve health risks for consumers, as the undeclared alterations in chemical composition could cause potential allergic effects [5]. Overall, adulteration reduces the product's quality, deceiving consumers and violating their rights. Therefore, the monitoring of pomegranate juice authenticity by both regulatory agencies and the fruit juice industry is utterly essential [7].

In this context, the development of reliable, sensitive, and efficient analytical methodologies to detect pomegranate juice adulteration represents a demanding and challenging task. Conventional analytical techniques can be used to detect severe adulteration practices through the measurement of selected physicochemical indicators (pH, °Brix value, or titratable acidity), but they are often unable to detect small differences that could be indicative of low-level adulteration [8]. The most established approaches so far are based on targeted profiling of specific fruit juice constituents such as amino acids [2,9], polyphenols [4,10,11], and organic acids [2,6,12]. Based on this approach, Zhang et al. (2009) used a combination of existing databases and analytical techniques to characterize pure pomegranate juices and to establish authentication criteria by developing an international multidimensional authenticity specification (IMAS) algorithm [13]. Untargeted fingerprinting methodologies using spectroscopic techniques (Ultraviolet-visible spectroscopy, UV-VIS and Fourier-transform infrared spectroscopy, FTIR) combined with chemometrics have also been developed to unmask pomegranate juice adulteration through water addition or juice-to-juice adulteration with apple and grape juice [5,14].

In the last few years, a universal analytical approach called "metabolomics" has experienced a significant increase in interest in food fingerprinting studies [15,16]. Metabolomics focuses on the study of low-molecular-weight molecules (<1000 Da) and is used to explore and characterize food constituents, generating a detailed and comprehensive metabolic chemical profile of food. Metabolomic studies mainly involve the detection of metabolites (biomarkers) that can discriminate between sample populations (discriminative metabolomics) and/or the generation of statistical models able to classify samples and predict class memberships (predictive metabolomics) [17]. The identification and quantification of the biomarkers responsible for discrimination (informative metabolomics) is desirable but is not the main target in such studies [16]. In metabolomics, the use of high-throughput analytical techniques, such as high-resolution mass spectrometry (HRMS), is essential in order to enable the large-scale determination of unknown compounds. Statistical treatment using advanced chemometric tools, such as principal component analysis (PCA) and partial least squares discriminant analysis (PLS-DA), is necessary for the discrimination/classification of the samples and the development of predictive models [18].

Metabolomic studies have found, so far, limited application in fruit juice authenticity assessment, mainly concerning the detection of juice-to-juice adulteration of citrus fruits, while research regarding pomegranate juice adulteration is still scarce [8,19–22]. The main objective of this work was to evaluate the feasibility of targeted and untargeted Mass Spectrometry (MS)-based metabolomics, using ultraperformance liquid chromatography coupled to quadrupole time-of-flight mass spectrometry (UPLC-QTOF/MS), in the discrimination of authentic pomegranate juice and pomegranate juice adulterated with apple and red grape juice. The potential of this approach to detect low levels of juice-to-juice adulteration (down to 1%) was investigated, using both supervised (PLS-DA) and unsupervised (PCA) pattern recognition techniques. Finally, fingerprint compounds of apple and red grape juice were identified and structurally characterized and could be used as specific markers revealing pomegranate juice adulteration.

2. Materials and Methods

2.1. Chemicals and Reagents

All standards and reagents used were of high-purity grade (>95%): 2,5-dihydroxybenzoic acid (gentistic acid), 3,4-dihydroxybenzoic acid (protocatechuic acid), 4-hydroxybenzoic acid, 8-prenylnaringenin, catechin, chrysin, cinnamic acid, gallic acid, ferulic acid, epicatechin, *p*-coumaric acid, quercetin, vanillic acid, pinoresinol, syringaldehyde, syringic acid, taxifolin, salicylic acid, rutin, rosmarinic acid, resveratrol, pinobanksin, pinocembrin, myricetin, and eriodictyol were obtained from Sigma-Aldrich (Stenheim, Germany). Luteolin, hydroxytyrosol, and 2′,4′-dihydroxychalcone were purchased from Santa Cruz Biotechnology (Santa Cruz, CA, USA), while tyrosol, caffeic acid, vanillin, ethyl vanillin, apigenin galangin, genistein, hesperetin, and naringenin were purchased from Alfa Aesar (Karlsruche, Germany).

Methanol (MeOH) (LC–MS grade) was purchased from Merck (Darmstadt, Germany), whereas 2-propanol (LC–MS grade) was purchased from Fisher Scientific (Geel, Belgium). Sodium hydroxide monohydrate for trace analysis ≥99.9995%, ammonium acetate, and formic acid 99% were purchased from Fluka (Buchs, Switzerland). Distilled water was provided by a Milli-Q purification apparatus (Millipore Direct-Q UV, Bedford, MA, USA). Finally, regenerated cellulose syringe filters (RC filters, pore size 0.2 μm, diameter 15 mm) were acquired from Phenomenex (Torrance, CA, USA). Stock standard solutions of individual compounds (1000 μg mL^{-1}) were prepared in MeOH and stored at −20 °C in amber glass bottles to prevent photodegradation. Working mix solutions of concentrations from 0.25 to 10 mg/L for each analyte were prepared by gradient dilution of the stock solutions in methanol/water (50:50 *v/v*).

2.2. Samples and Sample Preparation

Twenty-eight commercial, concentrated fruit juice samples (five 100% pomegranate juices, eight 100% apple juices, and 15 100% red grape juices), were directly supplied by a major Greek fruit juice company, DELTA FOODS S.A (Athens, Greece) (Table 1). With Turkey being an important and growing player in the pomegranate market, commercial pomegranate juices belonging to the Turkish Hicaz variety were selected for this study. Hicaz is the most produced and most consumed pomegranate variety in Turkey and is widely exported to European countries [23]. Apple juice samples included two apple cultivars (Starking and Granny Smith) from three different geographical regions of Greece (Western Macedonia, Central Macedonia, and Thessaly), and red grape juices consisted of pool samples mixing seven varieties (Sangiovese, Montepulciano, Lambrusco, Schiava, Shiraz, Ciliegiolo, and Merlot), which came from Italy (Puglia). All concentrated juice samples were produced in 2016 and were diluted to 11.2 ± 0.5 °Brix for apple juice, 15 ± 0.5 °Brix for pomegranate juice, and 15.9 ± 0.5 °Brix for red grape juice prior to analysis, according to the manufacturer's instructions. Additionally, one freshly squeezed pomegranate juice was prepared from Ermioni variety fruits, hand-picked from an orchard in Argolida, Greece. The juice was prepared based on the sampling methodology of Arbona et al. [24]: At least eight fruits, two from each direction on the pomegranate tree, were collected from 10 replicate trees (*n* = 100), and their juice was extracted through manual squeezing. Commercial and freshly squeezed fruit juice aliquots were stored at −20 °C until analysis, with no further processing. Right before LC-QTOF/MS analysis, the samples were thawed at room temperature, centrifuged, and filtered through regenerated cellulose (RC) syringe filters. To simulate adulteration, pomegranate juice admixtures with apple and red grape juice were constructed at 1%, 2%, 3%, 5%, 10%, and 20% adulteration. Separate adulterated samples were obtained from the Hicaz and Ermioni pomegranate varieties. Pool juice samples of each fruit were used for the adulteration experiments, prepared by mixing equal portions of individual pomegranate, apple, and red grape samples. All samples were analyzed in triplicate, and the average retention times and peak areas were calculated for each compound.

Table 1. Fruit juice samples.

Juice	Variety	Origin	°Brix	Sample Code
Apple	Starkin, Granny Smith	Greece (Pella, Imathia, Kastoria, Larissa)	11.2 ± 0.5	A1–A8
Red grape	Sangiovese, Montepulciano, Lambrusco, Schiava, Shiraz, Ciliegiolo, Merlot	Italy (Puglia)	15.9 ± 0.5	G1–G15
Pomegranate	Hicaz	Turkey	15 ± 0.5	P1–P5
	Ermioni	Greece (Argolida)	15.3	P6

Besides the pure and adulterated fruit juice samples, a quality control (QC) sample was also prepared and analyzed periodically throughout the batch to evaluate and ensure adequate analytical performance. The QC sample was constructed by mixing same-volume aliquots of all examined pure juices, representing both the sample matrix and metabolite composition of the samples. It was injected at the beginning of the LC-QTOF/MS analysis (six times for conditioning) and also at regular intervals (every 12 injections) to monitor potential instrumental drifts. Three exact mass retention time (EMRT) pairs (*m/z* 191.0516_1.3 min, *m/z* 489.1973_5.3 min, and *m/z* 304.1924_10.0 min) were monitored in terms of peak area and retention time (RT) stability, and in all cases the %relative standard deviations (RSDs) were below 15%.

2.3. LC-QTOF/MS Analysis

The analysis of fruit juices was carried out using an ultrahigh-performance liquid chromatography (UHPLC) system with a HPG-3400 pump (Dionex Ultimate 3000 RSLC, Thermo Fischer Scientific, Dreieich, Germany) coupled to a QTOF mass spectrometer (Maxis Impact, Bruker Daltonics, Bremen, Germany). Chromatographic separation was performed using an Acclaim RSLC C18 column (2.1 × 100 mm, 2.2 μm) from Thermo Fischer Scientific (Dreieich, Germany) preceded by a C18 guard column thermostatted at 30 °C. The mobile phase consisted of water/methanol (90:10 *v/v*, solvent A) and methanol (solvent B), both containing 5 mM of ammonium acetate, and the gradient elution program started with 1% B (flow rate of 0.2 mL min^{-1}) for 1 min, which was increased to 39% in 2 min and then to 99.9% (flow rate of 0.4 mL min^{-1}) in another 11 min. Here, 99.9% of B was kept constant for 2 min (flow rate of 0.48 mL min^{-1}), and then re-equilibration of the column was performed, restoring the initial conditions for 3 min. The injection volume was set up to 5 μL. Ionization was performed using an electrospray ionization interface (ESI), operating in negative mode, with the following operation parameters: A capillary voltage of 3500 V, a nebulizer gas pressure of 2 bar (N_2), drying gas at 8 L min^{-1}, an end-plate offset of 500 V, and a dry temperature of 200 °C.

For each sample, the full scan mass spectra were obtained in a range of 50–1000 *m/z* using Bruker broadband collision-induced dissociation (bbCID) mode. The Bruker bbCID function offers MS and MS/MS spectra within the same injection, with a scan rate of 2 Hz working at two different collision energies (CEs), one low (4 eV) and one high (25 eV). This mode provides high MS sensitivity, enabling the determination of even low-concentration marker compounds that can differentiate fruit juices and reveal juice-to-juice adulteration. However, the acquired MS/MS spectra were noisy and not compound-specific, rendering the identification of compounds rather difficult. For this reason, a second MS analysis was performed in AutoMS (data-dependent) acquisition mode. In AutoMS, the five most abundant ions per MS scan are selected and fragmented, and the applied collision energy is set to predefined values based on the mass and charge state of the ions. This mode provided clear and compound-specific MS/MS spectra, which were used for the structure elucidation of unknown marker compounds. For low-concentration marker compounds that were not within the five most abundant ions per MS scan and where no MS/MS spectra were obtained, a third MS analysis was performed using a preselected inclusion mass list containing the precursor ions of interest (exact masses). The fragmentation of these *m/z* was triggered when their MS spectra intensity exceeded a specific intensity threshold.

A QTOF external calibration was performed daily using sodium formate in a mixture of water:isopropanol (50:50 *v/v*), and also internal calibration was performed by calibrant injection at the beginning of each run (1st segment, 0.1–0.25 min). A typical resolving power (Full width at half maximum, FWHM) between 36,000 and 40,000 at *m/z* 226.1593, 430.9137, and 702.8636 was provided. The TASQ 1.4 and Data Analysis 4.1 Bruker Daltonics software packages (Bremen, Germany) were used for mass spectra interpretation and data processing.

2.4. Screening Strategies

2.4.1. Target Screening

Among fruit secondary metabolites, phenolic compounds constitute a wide class of biomarkers, and their study has proven to be a powerful tool for assessing fruit juice authentication. Phenolic profiling has provided very promising results concerning the detection of juice-to-juice adulteration, as specific variations in a juice's phenolic profile can confirm which fruits are present [25,26]. Thus, our study was targeted mainly at the detection and identification of unique phenolic compounds that could serve as markers for the presence of apple and red grape juice in pomegranate juice.

A target database was built that included 37 phenolic compounds from different classes (flavones, flavonols, flavanols, flavanones, and phenolic acids) for which reference standards were commercially available. The database included information on the analytes' molecular formulas, pseudomolecular ions [M-H]⁻, retention times, and MS/MS fragments (qualifier ions) and is presented in Table S1 in "Electronic Supplementary Materials". Identification of the target compounds in the samples was performed on the basis of mass accuracy, isotopic fitting, retention time, and MS/MS fragments, and specific criteria thresholds were set. The mass error should not have exceeded 2 mDa for both the precursor ion and the qualifier ions, while mSigma values, measuring the isotopic fitting between the measured and theoretical molecular formulas, should have been below or equal to 50. The retention time tolerance threshold was set at ±0.2 min, the minimum peak area threshold at 800, and the minimum intensity threshold at 200, as reported in a previous study by our group [27]. Quantification of the analytes in pure and adulterated juice samples was performed through an external standard calibration method using standard solution calibration curves.

The developed LC-QTOF/MS target methodology was validated in order to verify its suitability for identification and quantification purposes. The validation was performed using pomegranate juice samples spiked with different concentrations of the targeted compounds. Linearity was evaluated using standard solutions, prepared as described in Section 2.1, and the intraday precision of the analyses was calculated by analyzing six replicates of spiked pomegranate samples at a concentration level of 5 mg/L. The method limits of detection (MLODs) and method limits of quantification (MLOQs) were defined as the analyte's concentration at which the signal-to-noise ratio (S/N) was above 3 and 10, respectively, and the matrix effect was evaluated by comparing standard solutions of the analytes prepared in pure solvent and in pomegranate juice samples according to the following equation:

$$\%\text{Matrix Effect} = ((\text{Peak area matrix matched standard/Peak area standard in pure solvent}) - 1) \times 100. \quad (1)$$

2.4.2. Nontarget Screening

Initially, LC-HRMS raw data files of all 29 samples analyzed were converted to mzXML files using ProteoWizard open source software (Proteowizard, Palo Alto, CA, USA). These files were transferred to the R environment and processed with an XCMS package using the centWave method for peak picking. The CentWave feature detection algorithm has been successfully used for LC-HRMS data, directly detecting regions of interest (ROIs) in the *m/z* domain [28]. XCMS peak picking parameters such as tolerated mass deviation ("ppm") and minimum and maximum chromatographic peak width ("min peakwidth, max peakwidth") were optimized using the IPO package in the R environment [29], and the optimized parameter results were 23.3, 17.5, and 40, respectively. The chromatographic signal-to-noise threshold ("snthresh" parameter) was set at a default value of 3 to filter noisy peaks. A prefilter (intensity filter defined as the threshold for an m/z to be considered a peak appearing in k consecutive scans at J intensity threshold (k,J)) was adjusted at (31,000) to discard false peaks early in the detected ROIs. A retention time correction was performed using a nonlinear retention time alignment wrapping algorithm through loess, and a final step of filling in the missing peaks was implemented to replace the missing values of nondetected peaks with a small value of the intensity [30]. Finally, the CAMERA and Non-target R packages were used complementarily for the annotation of isotope and adduct peaks [31,32].

After peak picking, a differential analysis was performed between authentic pomegranate, apple, and red grape fruit juices, which were processed in pairs. Nonparametric independent (unpaired) two-group tests (one-way analysis of variance (ANOVA) and Welch's t-test) were used to find the mass features (including accurate mass values and retention times) that differentiated pomegranate juice from apple juice and pomegranate juice from red grape juice. An unpaired differential analysis was selected because the two authentic juices were expected to have different chemical profiles (peak area or intensity measurements of detected compounds) and there was no knowledge about the parameters of data distribution between two groups [33]. In general, two-group tests allow for the determination of metabolite features whose levels are significantly different between two sets of samples. Here, fold changes (variations in the maximum intensity of m/z values at a given retention time between two groups), p-values (to filter in the m/z values whose intensity/peak area changes were significant between two groups), and Welch's t-test (to derive the group-regulated data for each m/z) were used [34,35].

Following the application of a nontarget screening workflow, a large dataset consisting of mass features (including accurate mass values and retention times) that discriminated pomegranate from apple and red grape juice samples was obtained. Mass features that were detected in apple and red grape but not in pomegranate juice or that presented great differences in abundance (more than 5 times higher abundance in apple and grape juice) were selected as m/z markers of interest, as they could reveal potential pomegranate juice adulteration. Two suspect databases were compiled, including the mass features of interest for each adulterant, and pomegranate–apple and pomegranate–grape adulterated samples were screened accordingly. The mass features that were determined in the adulterated pomegranate juices presented unique authenticity markers, unmasking pomegranate juice adulteration at different adulteration levels.

Selected m/z markers were tentatively identified according to their mass accuracy (<5 mDa), isotopic fit, MS/MS fragmentation pattern, and retention time. Elemental compositions of precursors and fragment ions were proposed, and probable molecular formulas were suggested using the Bruker Smart Formula Manually tool in Data Analysis 4.1. MS/MS spectra were examined and interpreted using literature data, spectral libraries such as MassBank [36], an online database search (FoodB, METLIN, and CHEBI), and in silico fragmentation tools, mainly Metfrag [37].

2.5. Chemometric Analysis

A multivariate statistical analysis was performed using unsupervised and supervised pattern chemometric techniques (PCA and PLS-DA) through an in-house program called ChemoTrAMS [38] in the R environment (RStudio, Version 1.1.463, Boston, MA, USA). A PCA was applied to the data obtained from target analysis and was used to locate any existing clustering of fruit juices based on their composition (pomegranate, apple, and grape juices) and their authenticity (pure or adulterated pomegranate juices). A PCA was used as an initial descriptive approach, while PLS-DA, as a supervised method, was applied to construct the supervised classification and prediction models. For this purpose, a dataset was constituted that included the variables (markers) obtained from nontarget screening in both authentic and adulterated pomegranate juice samples. The autoscaling method was used to remove any variation comprised during analysis (such as a loss of instrumental sensitivity) of an original HRMS peaks list. PLS-DA models were built and were able to determine the percentage of adulteration in pomegranate fruit juices and also the adulterant (apple or red grape). The reliability of the classification models was studied in terms of goodness-of-fit (R^2, recognition ability) and goodness-of-prediction (Q^2, prediction ability).

3. Results and Discussion

3.1. Target Screening

The developed targeted LC-QTOF/MS methodology was used to screen authentic and adulterated pomegranate, apple, and red grape fruit juices. Eighteen compounds were determined: Six phenolic acids (gentistic acid, caffeic acid, cinnamic acid, ferulic acid, *p*-coumaric acid, salicylic acid), eight flavonoids (epicatechin, eriodictyol, myricetin, naringenin, quercetin, taxifolin, catechin, rutin), two phenolic alcohols (tyrosol and hydroxytyrosol), one stilbenoid (resveratrol), and one phenolic aldehyde (syringaldehyde). For all of the determined compounds, the mass accuracies of both precursor ions and qualifier ions were <2 mDa compared to standard solutions, and also the mSigma value (isotopic fit) was <50. Quantification of the compounds was performed using their corresponding standard solution calibration curves, and the concentrations of the phenolic compounds determined were calculated as the average value ± the standard deviation of triplicate analyses for each sample. In every case, the %RSD of the three replicates did not exceed 10% for each individual sample. The target screening results for the authentic fruit juices examined are presented in Table 2.

Table 2. Concentrations of phenolic compounds in pure pomegranate, apple, and red grape juices. LOQ: Limit of quantification.

Compound	Pomegranate Juice, Hicaz Variety ($n = 5$) Concentration Range (mg/L)	Pomegranate Juice, Ermioni Variety ($n = 1$) Concentration (mg/L)	Apple Juice ($n = 8$) Concentration Range (mg/L)	Red Grape Juice ($n = 15$) Concentration Range (mg/L)
Caffeic acid	0.045–0.12	<LOQ	2.9–5.3	0.58–1.4
Catechin	0.71–1.1	4.1	0.94–1.1	12.2–46.3
Cinnamic acid	0.36–0.55	<LOQ	<LOQ	<LOQ
Epicatechin	0.039–0.083	2.2	4.0–8.2	4.4–14
Eriodictyol	0.12–0.18	<LOQ	0.35–0.42	0.10–0.24
Ferulic acid	0.43–0.78	<LOQ	0.19	0.27–0.90
Gentistic acid	1.0–1.6	2.9	0.49–0.61	3.0–5.3
Hydroxytyrosol	<LOQ	<LOQ	<LOQ	2.3–4.4
Myricetin	0.32–0.45	0.24	<LOQ	0.20–0.60
Naringenin	0.20–0.32	0.28	<LOQ	0.16–0.36
p-coumaric acid	0.25–0.50	<LOQ	0.30–0.62	0.55–1.3
Quercetin	0.12–0.20	<LOQ	0.031–0.09	0.15–0.43
Resveratrol	<LOQ	<LOQ	<LOQ	0.17–1.09
Rutin	0.29–0.53	LOQ	1.1–2.5	<LOQ
Salicylic acid	<LOQ	0.16	<LOQ	0.56–2.4
Syringaldeyde	0.32–0.44	0.59	<LOQ	<LOQ
Taxifolin	0.032–0.060	0.054	0.040–0.060	0.32–0.84
Tyrosol	0.10–0.18	<LOQ	0.21–0.44	0.45–0.93

The most abundant polyphenolic compounds in apple juices were found to be epicatechin and caffeic acid, with the contents of rutin and catechin being relatively lower. These results were in agreement with previous reported results [39–41]. Epicatechin and caffeic acid were detected in significantly lower amounts in Hicaz pomegranate juice, suggesting that they could be used as potential markers to differentiate the two juices and reveal Hicaz pomegranate juice adulteration from apple juice. Indeed, adulteration experiments showed that the presence of epicatechin at a concentration ≥0.25 mg/L (three times higher than in authentic samples) was indicative of pomegranate juice adulteration corresponding to the addition of at least 3% apple juice. The same applied for caffeic acid, for which a concentration ≥0.36 mg/L indicated apple juice addition to pomegranate juice of 5% or more. The concentrations of both epicatechin and caffeic acid were found to have a linear correlation with the percentage of apple juice added to pomegranate juice. These results are presented in Figure 1. However, epicatechin could not be used as a marker to detect apple juice addition in the Ermioni variety of pomegranate juice, as this variety presented a significant amount of epicatechin (2.2 mg/L). Ermioni pomegranate juice debasing with apple juice could be revealed by the presence of caffeic acid at a level as low as 1%, since no caffeic acid was detected in the Ermioni variety of pomegranate juice (Figure S1).

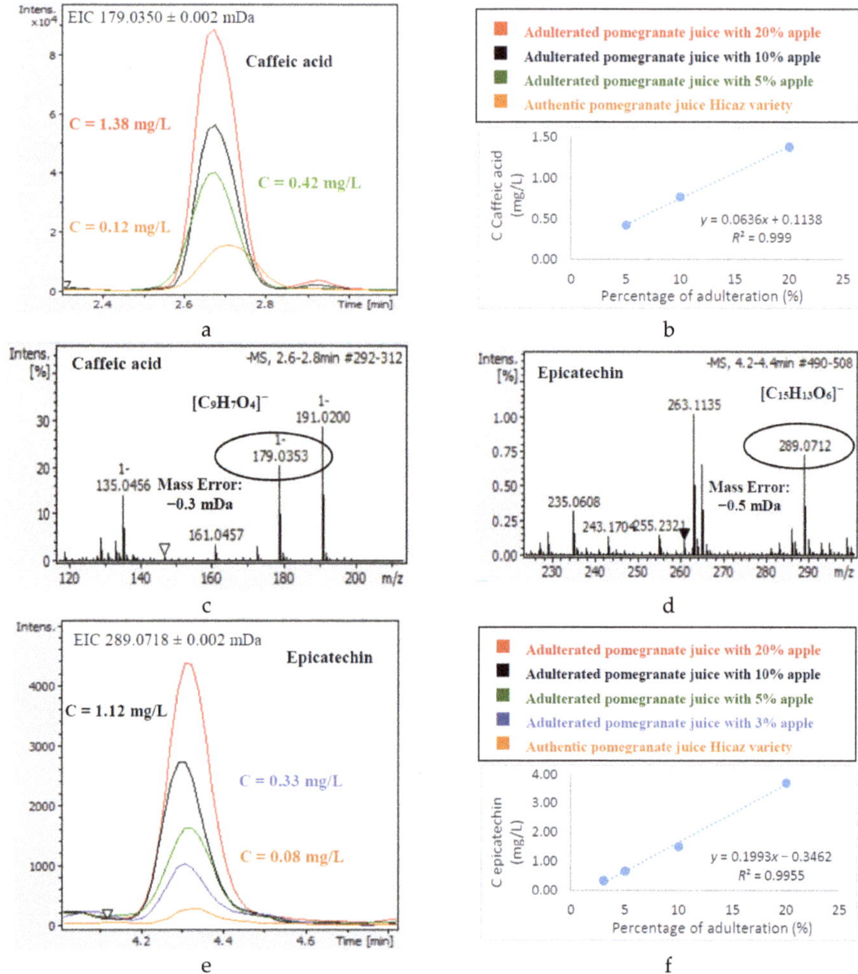

Figure 1. Extracted Ion Chromatograms and Mass Spectrometry (MS) spectra of epicatechin and caffeic acid in authentic and adulterated Hicaz pomegranate juices. (**a**) EIC of caffeic acid in authentic and adulterated pomegranate juice. (**b**) Linear regression of caffeic acid and adulteration percentage. (**c**) MS spectra of caffeic acid. (**d**) MS spectra of epicatechin. (**e**) EIC of epicatechin in authentic and adulterated pomegranate juice. (**f**) Linear regression of epicatechin and adulteration percentage.

Pomegranate juice adulteration from red grape juice could be detected based on the concentrations of epicatechin, catechin, hydroxytyrosol, and resveratrol. The presence of epicatechin in Hicaz pomegranate juice could reveal adulteration from red grape juice as low as 2% (at concentrations ≥0.25 mg/L), while catechin was a relatively less indicative marker, as it also exists in pomegranate juice in lower quantities (adulteration ≥5%). However, neither catechin nor epicatechin could be used as adulteration markers in Ermioni pomegranate juices, as they both exist in high amounts in this variety (Figure S2). Hydroxytyrosol proved to be a characteristic marker of pomegranate adulteration from red grape juice, as it could disclose 3% adulteration or higher (concentration (C) ≥ 0.35 mg/L) in both examined pomegranate varieties, which also applied for resveratrol (20% adulteration, C ≥ 0.15 mg/L) (Figure S3).

The performance of the developed UPLC-QTOF/MS target method was validated to ensure its suitability for identification and quantification purposes. Various analytical parameters were examined, including accuracy (recovery), precision (%RSD), limits of detection and quantification (LODs and LOQs), linearity (calibration curves) and matrix effects, the results for which are presented in Table S2 in the Electronic Supplementary Materials. Intraday precision was assessed in terms of the %RSD, which varied from 0.92% (chrysin) to 7.25% (salicylic acid) and proved the excellent repeatability of the proposed methodology. Calibration curves were constructed in a concentration range from 0.25 to 10 mg/L, displaying excellent linearity with correlation coefficients >0.99 for all analytes. All target compounds showed adequate recovery efficiency (58.8% for gallic acid to 103% for 2′,4′-dihydroxychalcone), and relatively low matrix effects were observed, with 30 out of 37 compounds presenting matrix effects ±40%. The LODs and LOQs were satisfactory, ranging between 0.0095 mg/L (hesperetin) and 0.087 mg/L (catechin) and 0.029 and 0.26 mg/L, respectively.

3.2. Nontarget Screening

The application of the nontarget screening workflow and the differential analysis in pomegranate and apple juice samples produced 1054 m/z markers that were detected in apple juice but not in pomegranate juice (or demonstrated a great difference in abundance). A fold value threshold of 10 was applied in order to distinguish the most robust and reliable markers responsible for differentiating the fruit juices, and 214 m/z markers were further investigated to evaluate their usefulness as adulteration markers. An in-house suspect list that included these m/z markers was built, and all authentic and adulterated pomegranate juice samples were screened using TASQ 1.4 from Bruker. Four out of 214 important mass features already existed in the target list (epicatechin, catechin, caffeic acid, and rutin) and thus were excluded from the nontarget list. From the 214 mass features investigated, 67 could disclose 20% adulteration or more, as they exhibited at least three times higher abundance in adulterated samples compared to authentic ones. Similarly, 48 mass features were indicative of 10% adulteration, 28 of 5%, 14 of 3%, 27 of 2%, and 3 mass features could even reveal 1% adulteration of pomegranate juice from apple juice (Table S3). From the annotation of these 67 mass features, and after excluding adducts, isotopes, and in-source fragments, 42 marker compounds were determined to reveal pomegranate juice adulteration from apple juice. Additionally, five mass features were identified as double-charged compounds (m/z 588.1883_3.4 min, m/z 446.0816_1.0 min, m/z 728.2276_3.4 min, m/z 609.1929_3.3 min, and m/z 579.1475_1.0 min), as they revealed characteristic isotopic patterns.

For pomegranate–red grape juice samples, 1335 m/z markers characteristic of grape juice were obtained. Following the same procedure, 191 significant mass features were included in the second in-house suspect list, which was used to screen all authentic and adulterated pomegranate juice samples. Six out of these 191 mass features already existed in the target list (epicatechin, catechin, hydroxytyrosol, salicylic acid, taxifolin, and resveratrol). After a careful examination of the dataset of adulterated pomegranate juice samples from grape juice, 47 m/z markers were found to disclose adulteration at a level of 20% or more; 37 at a level of 10%; 17, 10, and 4 m/z markers at levels of 5%, 3%, and 2%, respectively; and, finally, 3 m/z markers were able to reveal pomegranate juice adulteration even at a level of 1%. Annotation of these mass features led to the determination of 45 marker compounds, as two mass features (m/z 163.0401_1.7 min and m/z 203.1076_6.6 min) were found to belong to in-source fragments of other marker compounds. All results are presented in Table S4 in "Electronic Supplementary Materials".

3.3. Tentative Identification of Marker Compounds

The identification of the characteristic markers that discriminated between adulterated and authentic pomegranate juices represented one of the most difficult and challenging steps of the metabolomics workflow. The use of LC-HRMS-QTOF instrumentation ensured the acquisition of accurate MS and MS/MS spectra, which were essential for the reliable elemental formula estimation of mass features of interest. The probable elemental compositions of marker compounds were computed using "SmartFormula Manually" from Bruker, which is based on accurate mass determination and isotopic patterns. C ($n \le 50$), H ($n \le 100$), O ($n \le 20$), N ($n \le 10$), and S ($n \le 5$) atoms were considered for the molecular formula calculations. The proposed formulas were sorted according to the SmartFormula Manually Score (the most probable proposed formula scored 100%). Subsequently, a stepwise search of the biomarkers' proposed molecular formulas (in a descending order) was performed in several online databases, such as MassBank (http://www.massbank.jp/?lang=en), METLIN (https://metlin. scripps.edu/landing_page.php?pgcontent=mainPage), ChEBI (https://www.ebi.ac.uk/chebi/), and FoodB (http://foodb.ca/), with the latter more focused on natural product constituents. Only candidates that could possibly be present in fruit juices were further examined, and the experimental MS/MS mass spectra were compared to those provided in the databases and/or literature. In silico fragmentation with Metfrag [37] was also performed to elucidate the chemical structure of potential biomarkers.

For pomegranate juice adulteration with apple juice, 22 out of 42 marker compounds were tentatively identified. Three EMRTs were found to reveal adulteration down to 1%: *m/z* 353.0879_2.9 min, *m/z* 191.0564_2.9 min, and *m/z* 193.0509_6.1 min. However, these mass features corresponded to only two different marker compounds, as the *m/z* 191.0564_2.9 min ion was proven to be an in-source fragment of *m/z* 353.0879_2.9 min (Figure 2a–c). For the mass feature *m/z* 353.0877_2.9 min, the most probable molecular formula was suggested to be $C_{16}H_{18}O_9$, with a mass error of 0.1 mDa and an mSigma value of 7.1 (Figure 2d). This molecular formula corresponded to 16 potential candidates in the FoodB and Metlin databases, from which only chlorogenic acid and its isomers have been reported to exist in fruits. The experimental MS/MS spectra obtained (Figure 2e) were compared to the MS/MS spectra of chlorogenic acid that are reported in MassBank (Figure 2f), and two common fragments were revealed (Figure 2g). A reference standard was then obtained for chlorogenic acid, and its presence in the samples was confirmed (Figure 2g–i). Following the same workflow, the mass feature with *m/z* 193.0509_6.1 min was tentatively identified as vanillin acetate (Figure S4 in "Electronic Supplementary Materials"). The probable elemental compositions and tentative identification of all marker compounds that revealed pomegranate juice adulteration from apple juice are presented in Table 3. Identification data for selected marker compounds are presented in "Electronic Supplementary Materials", Figures S4–S12.

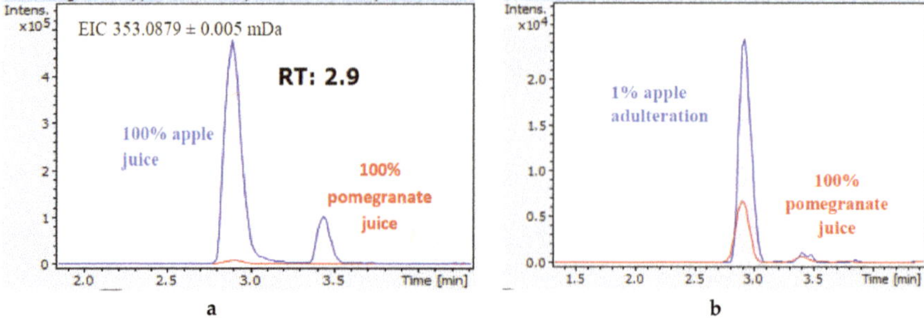

Figure 2. *Cont.*

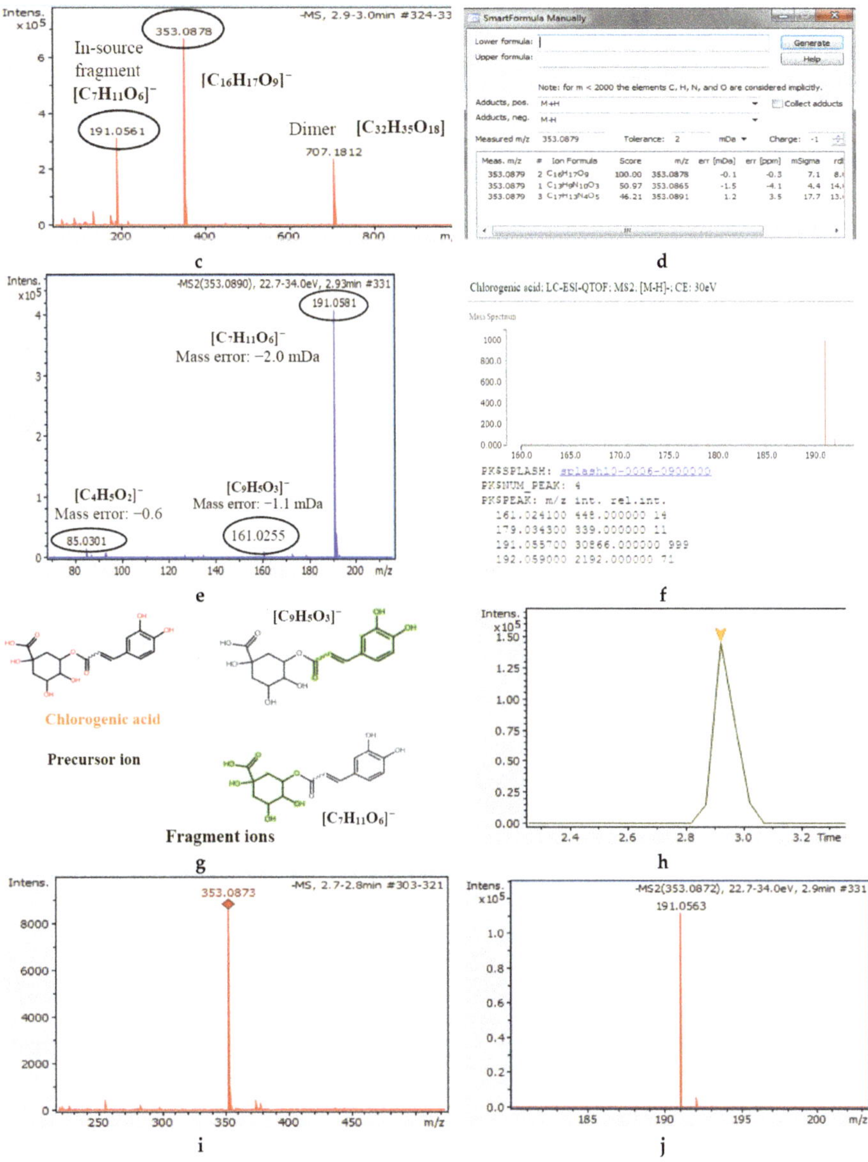

Figure 2. Identification data for the mass feature *m/z* 353.0878_2.9 min (chlorogenic acid). (**a**) EIC of *m/z* 353.0879 in apple–pomegranate juice. (**b**) EIC of *m/z* 353.0879 in authentic and adulterated pomegranate juice. (**c**) MS spectra of mass feature *m/z* 353.0879_2.9 min. (**d**) Probable elemental composition of mass feature *m/z* 353.0879_2.9 min. (**e**) MS/MS spectra of mass feature *m/z* 353.0879_2.9 min. (**f**) MS/MS spectra of chlorogenic acid (MassBank record FIO00625). (**g**) Precursor and fragment ions of chlorogenic acid. (**h**) EIC of *m/z* 353.0879 in the chlorogenic acid reference standard. (**i**) MS spectra of chlorogenic acid. (**j**) MS/MS spectra of chlorogenic acid.

Table 3. Tentative identification of characteristic marker compounds indicating pomegranate juice adulteration from apple juice.

# Marker	m/z (Precursor Ion)	Retention Time (min)	Ion	m/z (Fragment Ions)	Probable Elemental Composition	Mass Error (mDa)	Tentative Identification	Indicative Level of Adulteration
1	353.0879	2.9	[M-H]⁻	191.0564; 192.0611; 93.0345	$C_{16}H_{18}O_9$	0.1	Chlorogenic acid	1%
2	193.0509	6.1	[M-H]⁻	133.0289; 178.9993	$C_{10}H_{10}O_4$	0.3	Vanillin acetate	1%
3	353.0880	3.4	[M-H]⁻	191.0565	$C_{16}H_{18}O_9$	0.2	Chlorogenic acid isomer	2%
4	183.0664	3.7	[M-H]⁻	71.0141; 138.0560	$C_9H_{12}O_4$	0.1	Unknown compound	2%
5	337.0942	3.9	[M-H]⁻	173.0464; 163.0407; 119.0505	$C_{16}H_{18}O_8$	1.4	p-coumaroylquinic acid	2%
6	191.0551	1.3	[M-H]⁻	85.0302; 72.9937; 127.0411	$C_7H_{12}O_6$	1.00	Quinic acid	3%
7	273.0771	5.9	[M-H]⁻	167.0357; 123.0467; 125.0249	$C_{15}H_{14}O_5$	−0.3	Phloridzin (in-source fragment)	3%
8	307.1762	6.3	[M-H]⁻	161.0464; 71.0132	$C_{14}H_{28}O_7$	0.1	(R)-1-O-b-ᴅ-glucopyranosyl-1,3- octanediol	3%
9	351.1309	3.9	[M-H]⁻	101.0613; 249.0630; 291.1094	$C_{14}H_{24}O_{10}$	−1.2	2-O-acetyl-α-ᴅ-abequopyranosyl-(1→3)-α-ᴅ-mannopyranose	3%
10	161.0819	2.1	[M-H]⁻	130.088; 109.0297; 153.0215	$C_7H_{14}O_4$	−0.6	Unknown compound	3%
11	517.2284	4.5	[M-H]⁻	385.1865; 205.1236; 149.0457; 293.0879	$C_{24}H_{38}O_{12}$	0.7	Vomifoliol 9-[xylosyl-(1->6)-glucoside]	5%
12	165.0776	1.5	[M-H]⁻	89.0247; 119.0359; 149.047	$C_6H_{14}O_5$	−0.7	ʟ-rhamnitol	5%
13	195.0882	1.5	[M-H]⁻	71.0142; 59.0141; 73.0298	$C_7H_{16}O_6$	−0.4	Unknown compound	5%
14	289.0830	2.8	[M-H]⁻	245.0934; 203.0828; 116.0499	$C_{14}H_{14}N_2O_5$	0.0	N2-malonyl-ᴅ-tryptophan	5%
15	337.0942	3.3	[M-H]⁻	191.0576; 163.042; 119.0512	$C_{16}H_{18}O_8$	−1.3	p-coumaroylquinic acid isomer	5%
16	405.1778	3.5	[M-H]⁻	225.1153; 181.1243; 71.0149	$C_{18}H_{30}O_{10}$	−1.3	Unknown compound	5%
17	393.1768	5	[M-H]⁻	125.0255; 161.0444; 249.1363	$C_{17}H_{30}O_{10}$	−0.2	Unknown compound	5%
18	425.167	3.9	[M-H]⁻	235.1203; 143.0386; 287.0537	$C_{17}H_{30}O_{12}$	−0.6	Unknown compound	5%
19	498.1270	5.9	[M-H]⁻	273.078; 167.0364; 307.1781	$C_{36}H_{19}O_3$	−0.9	Phloridzin-related compound	5%
20	351.1299	3.6	[M-H]⁻	191.0595; 71.0143; 101.0622	$C_{14}H_{24}O_{10}$	−1.2	Unknown compound	10%
21	451.1243	5.3	[M-H]⁻	289.0718; 167.0355; 125.0249	$C_{21}H_{24}O_{11}$	0.3	3-Hydroxyphloretin 2′-O-glucoside	10%
22	517.3162	7.4	[M-H]⁻	285.0398; 383.2585	$C_{30}H_{46}O_7$	−0.8	Corosin	10%
23	429.1769	3.7	[M-H]⁻	205.1250; 249.1141; 161.1353	$C_{20}H_{30}O_{10}$	−0.4	Unknown compound	10%
24	469.2284	5.8	[M-H]⁻	273.0772; 300.029; 433.0792;	$C_{20}H_{38}O_{12}$	−0.5	In-source fragment of compound with m/z 529.2496	10%
25	456.151	4.1	[M-H]⁻	145.0311; 133.0144;	$C_{20}H_{27}NO_{11}$	−0.1	Unknown compound	10%
26	497.2234	5.1	[M-CH₃COOH]⁻	305.1618; 131.0358;179.0577	$C_{21}H_{38}O_{13}$	−0.6	Ebracteatoside D	10%
27	567.1720	5.5	[M-H]⁻	273.0774; 167.0359; 125.0252	$C_{26}H_{32}O_{14}$	−0.1	Phloretin 2′-xyloglucoside	10%
28	510.0888	6.0	[M-H]⁻	300.0285; 301.0358; 447.0952	$C_{21}H_{21}NO_{14}$	0.1	Unknown compound	10%
29	439.2180	6.0	[M-H]⁻	307.1703	$C_{19}H_{36}O_{11}$	0.5	Unknown compound	10%
30	413.1306	1.5	[M-H]⁻	235.0521	$C_{15}H_{26}O_{13}$	−0.6	Unknown compound	20%
31	467.1191	2.0	[M-H]⁻	305.0717	$C_{21}H_{24}O_{12}$	−0.4	Unknown compound	20%
32	583.1663	4.9	[M-H]⁻	167.0354; 289.0717; 125.0240	$C_{26}H_{32}O_{15}$	0.5	3-hydroxyphloretin 2″-O-xylosylglucoside	20%
33	597.1814	5.2	[M-H]⁻	273.0796; 167.0361; 179.0388	$C_{27}H_{34}O_{15}$	1.1	Phloridzinyl glucoside	20%
34	485.2236	6.0	[M-H]⁻	59.0142; 71.0145	$C_{20}H_{38}O_{13}$	0.4	Unknown compound	20%
35	273.0766	7.5	[M-H]⁻	167.0357; 123.0467; 145.0355	$C_{15}H_{14}O_5$	0.2	Phloretin	20%
36	475.1313	1.4	[M-H]⁻	133.0141; 115.0034; 179.0564	$C_{32}H_{56}O_{32}$	−0.8	Unknown compound	20%
37	207.0652	7.0	[M-H]⁻	161.0272; 133.0292; 179.0381	$C_{11}H_{12}O_4$	1.0	Unknown compound	20%
38	337.1147	3.1	[M-H]⁻	249.0616; 87.0454; 175.0076	$C_{13}H_{22}O_{10}$	−0.7	2,2-bis[[3-hydroxy-2-(hydroxymethyl)-2-methyl-propanoyl]oxy]propanoic acid	20%
39	425.2025	5.7	[M-H]⁻	326.0673	$C_{18}H_{34}O_{11}$	0.3	Unknown compound	20%
40	463.0883	5.5	[M-H]⁻	300.0288; 271.0268; 151.0016	$C_{21}H_{20}O_{12}$	−0.1	Quercetin 3-galactoside	20%
41	501.3215	10.6	[M-H]⁻	483.3176; 409.3087; 483.3118	$C_{30}H_{46}O_6$	−0.6	Esculentic acid	20%
42	580.2237	4.5	[M-H]⁻	149.0467; 205.1241	$C_{21}H_{42}O_{18}$	−1.7	Unknown compound	20%

Phloridzin fragments with m/z 273.0771_5.9 min and phloretin (m/z 273.0766_7.5) were confirmed with reference standards. Although phloridzin is a unique apple juice marker, as is also reported in the literature [26,39], the adulteration of pomegranate juice from apple juice cannot be revealed by monitoring its precursor ion (m/z 435.0308). The reason is that an isobaric compound of phloridzin, tentatively identified as phenethyl 6-galloylglucoside, exists in pomegranate juice and is eluted at the same RT as phloridzin (Figure S5e). However, these two compounds can be easily distinguished by their different fragmentations (MS/MS spectra), as is shown in Figure S5d,g. Consequently, phloridzin's in-source fragment with m/z 273.0766 could be accurately used as an adulteration marker, revealing the presence of apple juice in pomegranate juice at a level down to 3% (Figure S5). Moreover, two of the most important markers, with m/z 373.0942, RT 3.9, and 3.3 min, were tentatively identified as *p*-coumaroylquinic acid isomers, detecting pomegranate juice adulteration of 2% and 5%, respectively. Both 4-*O*-*p*-coumarylquinic acid and 3-*O*-*p*-coumarylquinic acid have been reported to exist in apple juices, showing characteristic fragmentation patterns [26]. In the absence of reference standards to provide the exact RT and MS/MS spectra, we were unable to distinguish between positional isomers (Figure S6).

For pomegranate juice adulteration from grape juice, 18 out of 45 marker compounds were tentatively identified (Table 4). Three mass features were found to reveal adulteration down to 1%: m/z 369.0278_2.2 min, m/z 149.0096_1.2 min, and m/z 287.1502_4.0 min. The mass feature with m/z 149.0096_1.2 min was identified as tartaric acid, a well-known constituent of grape juice [2]. As presented in Figure 3, the molecular formula calculated by SmartFormula Manually for this mass feature was $C_4H_7O_6$, with a mass error of −0.5 mDa. Two characteristic fragment ions of tartaric acid were detected in the MS/MS spectra, $C_2HO_2^-$ with m/z 72.9932 and $C_3H_3O_3^-$ with m/z 87.0086, recording a score of 1.0 in MetFrag. Following the same workflow, 17 more compounds were identified, among them malvidin glucoside, resveratrol 3-glucoside, cis-coutaric acid, procyanidin B, quercetin 3-glucuronide, protocatechuic acid 4-glucoside, and peonidin 3-glucoside, as presented in Table 4 and Figures S13–S18 of the Electronic Supplementary Materials. Particularly for malvidin-3-*O*-glucoside, two characteristic ions of anthocyanin fragmentation in negative ionization were found in the MS spectra, [M-2H]⁻ with m/z 491.1191 and [M-2H + H_2O]⁻, according to a previous study of Sun et al. [42].

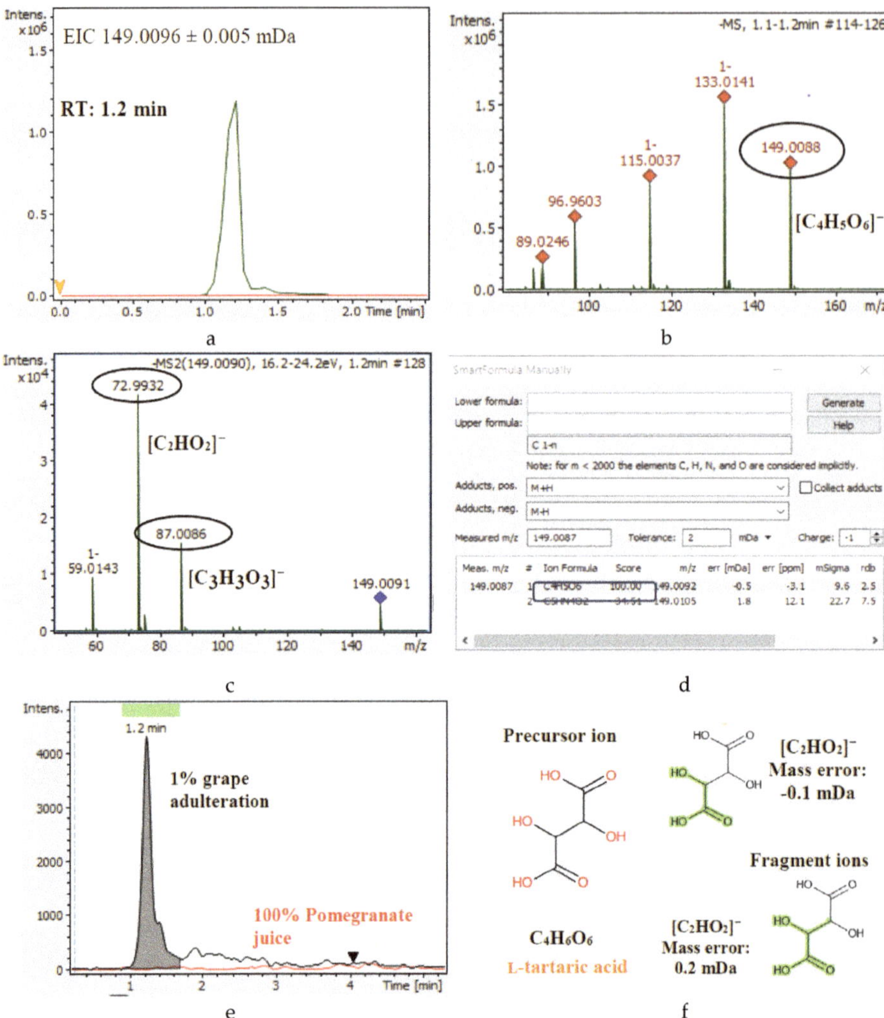

Figure 3. Identification data for the mass feature *m/z* 149.0096_1.2 min (L-tartaric acid). (**a**) EIC of *m/z* 149.0096 in pomegranate–grape juice. (**b**) MS spectra of mass feature *m/z* 149.0096_1.2 min. (**c**) MS/MS spectra of mass feature *m/z* 149.0096_1.2 min. (**d**) Probable elemental composition of mass feature *m/z* 149.0096_1.2 min. (**e**) EIC of *m/z* 149.0096 in authentic and adulterated pomegranate juice samples. (**f**) Structures of precursor and fragment ions of L-tartaric acid.

Table 4. Tentative identification of characteristic marker compounds indicating pomegranate juice adulteration from red grape juice.

# Marker	m/z (Precursor Ion)	Retention Time (min)	Ion	m/z (Fragment Ions)	Probable Elemental Composition	Mass Error (mDa)	Tentative Identification	Indicative Level of Adulteration
1	369.0278	2.2	[M-H]$^-$	125.0240; 161.0240; 80.9650	$C_{18}H_{10}O_9$	−2.6	Unknown compound	1%
2	149.0096	1.2	[M-H]$^-$	72.9932; 87.0086; 59.0143	$C_4H_6O_6$	0.4	ʟ-Tartaric acid	1%
3	287.1502	4.0	[M-H]$^-$	227.1283; 123.045; 203.0718	$C_{14}H_{24}O_6$	0.6	Unknown compound	1%
4	491.1191	4.7	[M-H]$^-$	328.0586; 329.0645; 313.0343; 330.0682	$C_{23}H_{25}O_{12}$	0.4	Malvidin-3-O-glucoside	2%
5	261.0405	4.9	[M-H]$^-$	125.0243; 61.9890; 197.0447; 204.1144	$C_{13}H_{10}O_6$	−0.2	Maclurin	3%
6	389.1242	4.8	[M-H]$^-$	227.0716; 185.0597; 143.0515	$C_{20}H_{22}O_8$	−0.9	Resveratrol 3-glucoside (cis-piceid)	3%
7	295.0464	1.7	[M-H]$^-$	163.0396; 119.0499; 87.0088	$C_{13}H_{12}O_8$	−0.4	cis-Coutaric acid	3%
8	283.0396	2.7	[M-H]$^-$	142.0659; 222.0223; 241.004	$C_{19}H_8O_3$	0.5	Unknown compound	3%
9	261.1344	3.0	[M-H]$^-$	73.0299; 187.0968; 201.1122	$C_{12}H_{22}O_6$	−0.1	Phaseolic acid	3%
10	311.0808	2.3	[M-H]$^-$	185.1174; 80.9647; 130.086; 229.1068	$C_{15}H_{12}N_4O_4$	−2.2	Unknown compound	3%
11	369.0278	3.0	[M-H]$^-$	125.0244;161.0242; 287.0564	$C_{18}H_{10}O_9$	−2.6	Unknown compound	3%
12	577.1346	3.3	[M-H]$^-$	289.0712; 125.0245; 407.0739; 161.0239; 245.0812	$C_{30}H_{26}O_{12}$	0.6	Procyanidin B isomer	3%
13	427.0340	4.1	[M-H]$^-$	347.0732; 165.0189; 261.0757	$C_{13}H_{16}O_{16}$	2.6	Unknown compound	5%
14	477.0671	5.0	[M-H]$^-$	301.0341; 151.0031; 178.9983; 316.0197	$C_{21}H_{18}O_{13}$	1.0	Quercetin 3-glucuronide	5%
15	509.1298	3.3	[M+H20-H]$^-$	149.0238; 329.0653; 193.0139; 165.0191; 347.0758	$C_{23}H_{26}O_{13}$	0.3	Quercetin 3,3′-dimethyl ether 4′-glucoside	5%
16	167.0348	5.5	[M-H]$^-$	123.0443; 81.0343	$C_8H_8O_4$	0.2	Unknown compound	5%
17	295.0858	3.8	[M-H]$^-$	169.1227; 80.9653; 213.1121; 170.1277	$C_{11}H_{20}O_7S$	−0.1	Unknown compound	5%
18	315.0725	2.0	[M-H]$^-$	152.0113; 255.2329; 217.0038	$C_{13}H_{16}O_9$	−0.3	Protocatechuic acid 4-glucoside	10%
19	121.0293	2.6	[M-H]$^-$	59.0144;66.0351	$C_7H_6O_2$	0.7	Benzoic acid	10%
20	397.0235	4.1	[M-H]$^-$	317.0653; 165.0190; 193.0141	$C_{12}H_{14}O_{15}$	2.5	Unknown compound	10%
21	295.0857	4.5	[M-H]$^-$	169.1229; 80.9649; 213.1120	$C_{15}H_{12}N_4O_3$	−2.1	Unknown compound	10%
22	461.1088	4.6	[M-H]$^-$	299.0550; 298.0482; 283.0248; 284.0319	$C_{22}H_{22}O_{11}$	0.1	Peonidin 3-glucoside	10%
23	231.1027	5.4	[M-H]$^-$	169.1018; 213.0925	$C_{14}H_{16}O_3$	0.00	Unknown compound	10%
24	219.1027	6.6	[M-H]$^-$	149.0956; 59.0149	$C_{13}H_{16}O_3$	0.00	Unknown compound	10%
25	637.1555	6.9	[M-H]$^-$	329.0658; 328.0569; 313.0351	$C_{32}H_{30}O_{14}$	0.8	Malvidin 3-(6-p-coumarylglucoside)	10%
26	423.0720	7.8	[M-H]$^-$	393.0249; 408.0440; 365.0305	$C_{22}H_{16}O_9$	0.1	Unknown compound	10%
27	591.1022	2.0	[M-H]$^-$	329.0654; 347.076; 411.0374	$C_{26}H_{24}O_{16}$	−3.0	Unknown compound	10%
28	446.0759	2.4	[M-H]$^-$	222.0219; 142.0658; 266.0106	$C_{20}H_{17}NO_{11}$	−3.0	Unknown compound	10%
29	369.0288	3.2	[M-H]$^-$	125.0239;161.0233; 165.0192	$C_{18}H_{10}O_9$	−2.6	Unknown compound	10%
30	190.0541	2.8	[M-H]$^-$	142.0463; 87.0080; 174.9927	$C_{10}H_9NO_3$	−3.1	Unknown compound	10%
31	577.1346	3.8	[M-H]$^-$	289.0705; 125.0241; 407.0739; 245.0812	$C_{30}H_{26}O_{12}$	0.6	Procyanidin B isomer	10%
32	305.0303	3.8	[M-H]$^-$	151.0041; 169.0143; 65.0031	$C_{14}H_{10}O_8$	0.1	Unknown compound	10%
33	161.0818	2.1	[M-H]$^-$	71.0516; 99.0847	$C_7H_{14}O_4$	0.2	Unknown compound	10%
34	209.0304	1.1	[M-H]$^-$	59.0139; 71.0142; 85.0347	$C_6H_{10}O_8$	0.2	Unknown compound	10%
35	293.1030	6.6	[M-H]$^-$	203.1071; 175.1126; 129.0550	$C_{15}H_{18}O_6$	0.1	Unknown compound	10%
36	429.2132	3.4	[M-H]$^-$	329.0667; 347.0770	$C_{21}H_{34}O_9$	−0.2	Unknown compound	10%
37	243.1239	3.6	[M-H]$^-$	61.9887; 73.0299; 125.0952;	$C_{12}H_{20}O_5$	−0.1	Unknown compound	20%
38	449.1087	5.3	[M-H]$^-$	151.0038; 285.0394; 178.9988	$C_{21}H_{22}O_{11}$	0.2	Taxifolin 3-rhamnoside	20%
39	131.0712	3.1	[M-H]$^-$	71.0140; 85.0654	$C_6H_{12}O_3$	0.2	Unknown compound	20%
40	330.2037	3.3	[M-H]$^-$	129.1035	$C_{15}H_{30}N_3O_5$	−0.3	Unknown compound	20%
41	366.1198	3.5	[M-H]$^-$	125.0976; 142.0668; 187.0979	$C_{17}H_{21}NO_8$	−0.4	Unknown compound	20%
42	107.0502	3.7	[M-H]$^-$	67.9611	C_7H_8O	0.00	Benzyl alcohol	20%
43	187.0974	3.3	[M-H]$^-$	125.0955; 57.0350; 123.0810	$C_9H_{16}O_4$	0.1	Azelaic acid	20%
44	373.1143	3.2	[M-H]$^-$	193.0504; 178.0269; 343.1009	$C_{16}H_{22}O_{10}$	−0.3	Geniposidic acid	20%
45	413.2403	5.3	[M-H]$^-$	169.0962	$C_{18}H_{38}O_{10}$	−1.1	Unknown compound	20%

3.4. Chemometric Analysis

Initially, the results of target screening methodology were used to differentiate authentic pomegranate, apple, and red grape juice samples. A PCA was performed on the 18 × 29 dataset (18 phenolic compounds were identified and quantified in 29 pure fruit juice samples). The PCA score plot generated for pure pomegranate, apple, and grape juice samples showed a distinctive separation between the three groups, with the first three principal components (PC1, PC2, and PC3) explaining the majority of the variation (58.8%, 25.5%, and 8.8%, respectively) (Figure 4). Subsequently, adulterated pomegranate samples with both apple and red grape juice were included with the existing PCA using the same dataset. However, the explained variance in PCs decreased, and there was no clear separation between adulterated and pure juice samples.

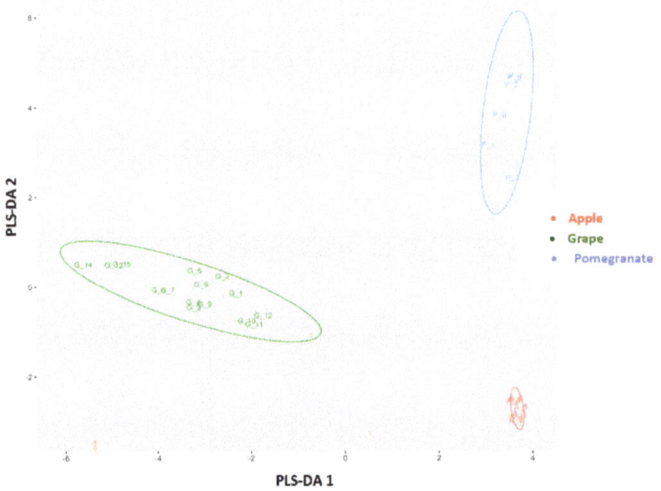

Figure 4. Principal component analysis (PCA) score plot showing the classification of authentic pomegranate, apple, and red grape juice samples.

PLS-DA was then applied to obtain classification models that could distinguish authentic pomegranate juices from adulterated ones in a supervised manner. At first, different models were built to differentiate pure pomegranate juices from those adulterated with apple and red grape juice. For the detection of pomegranate juices adulterated with apple juice, an autoscaled 28 × 51 dataset was used to build the model, in which rows represented the juice samples analyzed (28 objects) and columns the peak areas of the individual marker compounds, which were determined through both target and nontarget LC-QTOF/MS methodologies (51 variables). From the 28 authentic and adulterated juices, 18 juices were selected as the training set and 10 as the test set, randomly. The PLS-DA model correctly classified all authentic and adulterated pomegranate juice samples to a level of adulteration down to 1%. From the loading data, the most significant variables (using variable importance in projection (VIP) [27]) in the PLS-DA model were shown to be vanillin acetate and chlorogenic acid. Subsequently, a second model was built to detect pomegranate juice adulteration from red grape juice, again to a level of adulteration down to 1%. An autoscaled 28 × 50 dataset was used to build the second model (28 juice samples analyzed, 50 markers of red grape adulteration). Nineteen juices were selected in the training set and 9 in the test set, randomly. Again, the PLS-DA model successfully classified all authentic and adulterated pomegranate juice samples, even at the lowest level of adulteration (1%). From the loading data, the VIP variables that were indicative of adulteration from grape juice were hydroxytyrosol and an unknown mass feature with *m/z* 287.1502_4.0 min.

Finally, a third PLS-DA model was built to separate authentic pomegranate juice samples from adulterated ones containing either apple and/or red grape juice as the major adulterant. An autoscaled 52 × 39 dataset was used (42 samples, 39 variables) that included markers of adulteration with both fruit juices to a level down to 5% (Tables 3 and 4). The training set included 8 pure pomegranate juices (Hicaz cultivar) and also 30 juices adulterated with red grape and apple juice in a range of 20% to 1%. Six replicates of 1% juice adulteration (both red grape and apple) were included in the training set to increase the accuracy of the models at low adulteration levels. In order to evaluate the predictability of the model, the test set included 14 pure and adulterated pomegranate samples belonging to the Hicaz variety, but also authentic and adulterated freshly squeezed pomegranate juices from the Ermioni variety, prepared as described in Section 2.2. The prediction accuracy of the PLS-DA model was found to be more than adequate, as it successfully classified the authentic and adulterated pomegranate juices, even when they belonged to different varieties (Ermioni and Hicaz) (Figure 5). More specifically, the model successfully predicted all the adulteration ratios from red grape juice (down to 1%), while in the case of pomegranate juices adulterated with apple juice, it successfully predicted adulteration down to 2%, misclassifying only the 1% adulterated samples (they were indicated as pure pomegranate juice). The model cross-validation parameters were found to be very robust, with a goodness-of-fit (R^2) and goodness-of-prediction (Q^2) of 0.97 and 0.93, respectively, taking into consideration the first four PLS components. No outliers were observed according to Hotelling's $T2$ using a control limit of 95%. The results obtained through the PLS-DA models distinctly showed that the markers of adulteration that were detected through this study could be accurately used to achieve successful differentiation of authentic and adulterated pomegranate juices of different varieties.

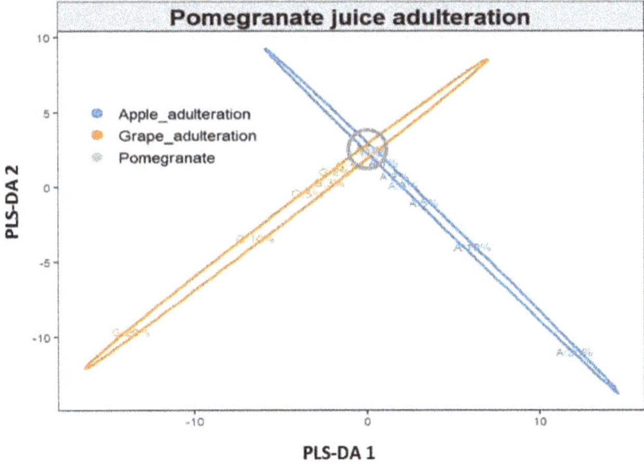

Figure 5. Partial least squares discriminant analysis (PLS-DA) plot showing the classification of authentic and adulterated pomegranate juice samples.

4. Conclusions

A novel approach was developed for the evaluation of pomegranate juice authenticity based on targeted and untargeted metabolomics coupled with advanced chemometric techniques. The combination of metabolomic profiling, metabolomic fingerprinting, and chemometrics provided a powerful approach for detecting pomegranate juice adulteration from apple and red grape juice at very low adulteration levels (down to 1%). The developed methodology is simple, sensitive, reliable, and robust and could be used not only for research purposes but also as an effective tool for the routine monitoring of pomegranate juice adulteration. To the best of our knowledge, this is the first study in the literature reporting more than 80 potential *m/z* markers that indicated the fraudulent addition of apple and/or grape juice in pure pomegranate juices in different portions. Several of these markers were identified, including phenolic acids, flavonoids, anthocyanins, and other minor metabolites. Processing of the mass spectrometric datasets of the m/z markers in authentic and artificially adulterated pomegranate samples by PCA and PLS-DA led to the construction of reliable and accurate classification and prediction models that could successfully discriminate between authentic and adulterated samples.

Supplementary Materials: The following are available online at http://www.mdpi.com/2304-8158/8/6/212/s1, Table S1: Target list of phenolic compounds, Table S2: Validation Data of target screening methodology, Table S3: Mass features revealing pomegranate adulteration with apple in different adulteration levels, Table S4: Mass features revealing pomegranate adulteration with grape in different adulteration levels, Figure S1: EICs and MS spectra of caffeic acid in authentic and adulterated Ermioni pomegranate juices, Figure S2: EICs and MS spectra of catechin and epicatechin in authentic and adulterated pomegranate juices, Figure S3: EICs and MS spectra of hydroxytyrosol and resveratrol in authentic and adulterated pomegranate juices, Figure S4: Identification data for the mass feature *m/z* 193.0509_6.1 min (vanillin acetate), Figure S5: Identification data for the mass feature *m/z* 273.0769_5.9 min (phloridzin in-source fragment), Figure S6: Identification data for the mass features *m/z* 337.0943_3.9 min and *m/z* 337.0943_3.3 min (*p*-coumaroylquinic acid isomers), Figure S7: Identification data for the mass feature *m/z* 191.0551_1.3 min (quinic acid), Figure S8: Identification data for the mass feature *m/z* 307.1762_6.3 min ((R)-1-*O*-b-ᴅ-glucopyranosyl-1,3-octanediol, Figure S9: Identification data for the mass feature *m/z* 351.1309_3.9 min (2-*O*-acetyl-alpha-ᴅ-abequopyranosyl-(1->3)-alpha-ᴅ-mannopyranose), Figure S10: Identification data for the mass feature *m/z* 517.2284 _4.5 min (Vomifoliol 9-[xylosyl-(1->6)-glucoside]), Figure S11: Identification data for the mass feature *m/z* 289.0830 _2.8 min (N2-malonyl-ᴅ-tryptophan); Figure S12: Identification data for the mass feature *m/z* 273.0766 _7.5 min (phloretin); Figure S13: Identification data for the mass feature *m/z* 491.1191_4.7 min (malvidin-3-*O*-glucoside), Figure S14: Identification data for the mass feature *m/z* 261.0403 _4.9 min (maclurin), Figure S15: Identification data for the mass feature *m/z* 389.1242 _4.8 min (resveratrol 3-glucoside), Figure S16: Identification data for the mass feature *m/z* 295.0464_1.7 min (cis-coutaric acid); Figure S17: Identification data for the mass feature *m/z* 261.1344_3.0 min (phaseolic acid); Figure S18: Identification data for the mass feature *m/z* 261.1344_3.0 min (procyanidin B).

Author Contributions: conceptualization, M.E.D. and N.S.T.; methodology, M.E.D. and R.A.; software, R.A. and S.K.D.; validation, M.E.D. and S.K.D.; data curation, M.E.D., S.K.D. and R.A.; writing—original draft preparation, M.E.D.; writing—review and editing, N.S.T.; project administration, M.E.D.; supervision, N.S.T.

Funding: Marilena Dasenaki's postdoctoral research was implemented under an IKY scholarship funded by the "Supporting Post-Doctoral Researchers" Act of the Operational Programme "Human Resources Development, Education and Lifelong Learning" with Priority Axes 6, 8, 9, which was cofunded by the European Social Fund (ESF) and Greek National Resources.

Acknowledgments: The authors are grateful to DELTA FOODS S.A. (23º klm Athens–Lamia, Ag. Stefanos–Attica) for its contribution to the supply of fruit juice samples from different locations in Greece and Europe.

Conflicts of Interest: The authors declare no conflicts of interest.

References

1. Global Fruit and Vegetable Juice Market Research Report. Available online: https://www.grandviewresearch.com/industry-analysis/fruit-vegetable-jui (accessed on 3 April 2019).

2. Nuncio-Jauregui, N.; Calin-Sanchez, A.; Hernandez, F.; Carbonell-Barrachina, A.A. Pomegranate juice adulteration by addition of grape or peach juices. *J. Sci. Food Agric.* **2014**, *94*, 646–655. [CrossRef] [PubMed]

3. Tastan, O.; Baysal, T. Adulteration Analysis of Pomegranate Juice. In *Frontiers in Drug Safety*; Shrivastava, A., Ed.; Bentham Science Publishers: Sharjah, UAE, 2018; Volume 1, pp. 91–100.

4. Borges, G.; Crozier, A. HPLC-PDA-MS fingerprinting to assess the authenticity of pomegranate beverages. *Food Chem.* **2012**, *135*, 1863–1867. [CrossRef] [PubMed]

5. Boggia, R.; Casolino, M.C.; Hysenaj, V.; Oliveri, P.; Zunin, P. A screening method based on UV-Visible spectroscopy and multivariate analysis to assess addition of filler juices and water to pomegranate juices. *Food Chem.* **2013**, *140*, 735–741. [CrossRef] [PubMed]

6. Dalmia, A. Rapid Measurement of Food Adulteration with Minimal Sample Preparation and No Chromatography Using Ambient Ionization Mass Spectrometry. *J. AOAC Int.* **2017**, *100*, 573–575. [CrossRef] [PubMed]

7. Zielinski, A.A.F.; Haminiuk, C.W.I.; Nunes, C.A.; Schnitzler, E.; van Ruth, S.M.; Granato, D. Chemical Composition, Sensory Properties, Provenance, and Bioactivity of Fruit Juices as Assessed by Chemometrics: A Critical Review and Guideline. *Compr. Rev. Food Sci. Food Saf.* **2014**, *13*, 300–316. [CrossRef]

8. Dasenaki, M.E.; Thomaidis, N.S. Quality and Authenticity Control of Fruit Juices—A Review. *Molecules* **2019**, *24*, 1014. [CrossRef] [PubMed]

9. Tezcan, F.; Uzasci, S.; Uyar, G.; Oztekin, N.; Erim, F.B. Determination of amino acids in pomegranate juices and fingerprint for adulteration with apple juices. *Food Chem.* **2013**, *141*, 1187–1191. [CrossRef] [PubMed]

10. Türkyılmaz, M. Anthocyanin and organic acid profiles of pomegranate (*Punica granatum* L.) juices from registered varieties in Turkey. *Int. J. Food Sci. Technol.* **2013**, *48*, 2086–2095. [CrossRef]

11. Spinelli, F.R.; Dutra, S.V.; Carnieli, G.; Leonardelli, S.; Drehmer, A.P.; Vanderlinde, R. Detection of addition of apple juice in purple grape juice. *Food Control* **2016**, *69*, 1–4. [CrossRef]

12. Ehling, S.; Cole, S. Analysis of organic acids in fruit juices by liquid chromatography-mass spectrometry: An enhanced tool for authenticity testing. *J. Agric. Food Chem.* **2011**, *59*, 2229–2234. [CrossRef]

13. Zhang, Y.; Krueger, D.; Durst, R.; Lee, R.; Wang, D.; Seeram, N.; Heber, D. International Multidimensional Authenticity Specification (IMAS) Algorithm for Detection of Commercial Pomegranate Juice Adulteration. *J. Agric. Food Chem.* **2009**, *57*, 2550–2557. [CrossRef] [PubMed]

14. Vardin, H.; Tay, A.; Ozen, B.; Mauer, L. Authentication of pomegranate juice concentrate using FTIR spectroscopy and chemometrics. *Food Chem.* **2008**, *108*, 742–748. [CrossRef] [PubMed]

15. Castro-Puyana, M.; Pérez-Míguez, R.; Montero, L.; Herrero, M. Application of mass spectrometry-based metabolomics approaches for food safety, quality and traceability. *TrAC Trend Anal. Chem.* **2017**, *93*, 102–118. [CrossRef]

16. Cubero-Leon, E.; Peñalver, R.; Maquet, A. Review on metabolomics for food authentication. *Food Res. Int.* **2014**, *60*, 95–107. [CrossRef]

17. Cevallos-Cevallos, J.M.; Reyes-De-Corcuera, J.I.; Etxeberria, E.; Danyluk, M.D.; Rodrick, G.E. Metabolomic analysis in food science: A review. *Trends Food Sci. Technol.* **2009**, *20*, 557–566. [CrossRef]

18. Hu, C.; Xu, G. Mass-spectrometry-based metabolomics analysis for foodomics. *TrAC Trend Anal. Chem.* **2013**, *52*, 36–46. [CrossRef]

19. Vaclavik, L.; Schreiber, A.; Lacina, O.; Cajka, T.; Hajslova, J. Liquid chromatography–mass spectrometry-based metabolomics for authenticity assessment of fruit juices. *Metabolomics* **2011**, *8*, 793–803. [CrossRef]

20. Twohig, M.; Krueger, D.A.; Gledhill, A.; Yang, J.; Burgess, J. Super fruit juice authenticity using multivariate data analysis, high resolution chromatography, UV and Time of Flight MS detection. *Agro Food Ind. Hi Tech* **2011**, *22*, 23–26.

21. Jandric, Z.; Roberts, D.; Rathor, M.N.; Abrahim, A.; Islam, M.; Cannavan, A. Assessment of fruit juice authenticity using UPLC-QToF MS: A metabolomics approach. *Food Chem.* **2014**, *148*, 7–17. [CrossRef]

22. Jandrić, Z.; Cannavan, A. An investigative study on differentiation of citrus fruit/fruit juices by UPLC-QToF MS and chemometrics. *Food Control* **2017**, *72*, 173–180. [CrossRef]

23. Turkish Exportal. Available online: https://www.turkishexportal.com/Pomegranate-Hicaz_SP8F3E_f909afe4630e4ef7b98eb6d8d17df7cd (accessed on 3 April 2019).

24. Arbona, V.; Iglesias, D.J.; Gomez-Cadenas, A. Non-targeted metabolite profiling of citrus juices as a tool for variety discrimination and metabolite flow analysis. *BMC Plant Biol.* **2015**, *15*, 38. [CrossRef] [PubMed]

25. Abad-García, B.; Garmón-Lobato, S.; Sánchez-Ilárduya, M.B.; Berrueta, L.A.; Gallo, B.; Vicente, F.; Alonso-Salces, R.M. Polyphenolic contents in Citrus fruit juices: Authenticity assessment. *Eur. Food Res. Technol.* **2014**, *238*, 803–818. [CrossRef]

26. Willems, J.L.; Low, N.H. Structural identification of compounds for use in the detection of juice-to-juice debasing between apple and pear juices. *Food Chem.* **2018**, *241*, 346–352. [CrossRef] [PubMed]

27. Kalogiouri, N.P.; Alygizakis, N.A.; Aalizadeh, R.; Thomaidis, N.S. Olive oil authenticity studies by target and nontarget LC-QTOF-MS combined with advanced chemometric techniques. *Anal. Bioanal. Chem.* **2016**, *408*, 7955–7970. [CrossRef] [PubMed]

28. Tautenhahn, R.; Bottcher, C.; Neumann, S. Highly sensitive feature detection for high resolution LC/MS. *BMC Bioinform.* **2008**, *9*, 504. [CrossRef] [PubMed]

29. Libiseller, G.; Dvorzak, M.; Kleb, U.; Gander, E.; Eisenberg, T.; Madeo, F.; Neumann, S.; Trausinger, G.; Sinner, F.; Pieber, T.; et al. IPO: A tool for automated optimization of XCMS parameters. *BMC Bioinform.* **2015**, *16*, 118. [CrossRef] [PubMed]

30. Smith, C.A.; Want, E.J.; O'Maille, G.; Abagyan, R.; Siuzdak, G. XCMS: Processing Mass Spectrometry Data for Metabolite Profiling Using Non-linear Peak Alignment, Matching, and Identification. *Anal. Chem.* **2006**, *78*, 779–787. [CrossRef] [PubMed]

31. Kuhl, C.; Tautenhahn, R.; Bottcher, C.; Larson, T.R.; Neumann, S. CAMERA: An integrated strategy for compound spectra extraction and annotation of liquid chromatography/mass spectrometry data sets. *Anal. Chem.* **2012**, *84*, 283–289. [CrossRef] [PubMed]

32. Loos, M.; Singer, H. Nontargeted homologue series extraction from hyphenated high resolution mass spectrometry data. *J. Cheminform.* **2017**, *9*, 12. [CrossRef]

33. Vinaixa, M.; Samino, S.; Saez, I.; Duran, J.; Guinovart, J.J.; Yanes, O. A Guideline to Univariate Statistical Analysis for LC/MS-Based Untargeted Metabolomics-Derived Data. *Metabolites* **2012**, *2*, 775–795. [CrossRef]

34. Tautenhahn, R.; Patti, G.J.; Rinehart, D.; Siuzdak, G. XCMS Online: A web-based platform to process untargeted metabolomic data. *Anal. Chem.* **2012**, *84*, 5035–5039. [CrossRef] [PubMed]

35. Gowda, H.; Ivanisevic, J.; Johnson, C.H.; Kurczy, M.E.; Benton, H.P.; Rinehart, D.; Nguyen, T.; Ray, J.; Kuehl, J.; Arevalo, B.; et al. Interactive XCMS Online: Simplifying advanced metabolomic data processing and subsequent statistical analyses. *Anal. Chem.* **2014**, *86*, 6931–6939. [CrossRef] [PubMed]

36. Horai, H.; Arita, M.; Kanaya, S.; Nihei, Y.; Ikeda, T.; Suwa, K.; Ojima, Y.; Tanaka, K.; Tanaka, S.; Aoshima, K.; et al. MassBank: A public repository for sharing mass spectral data for life sciences. *J. Mass Spectrom.* **2010**, *45*, 703–714. [CrossRef] [PubMed]

37. Wolf, S.; Schmidt, S.; Muller-Hannemann, M.; Neumann, S. In silico fragmentation for computer assisted identification of metabolite mass spectra. *BMC Bioinform.* **2010**, *11*, 148. [CrossRef] [PubMed]

38. Kalogiouri, N.P.; Aalizadeh, R.; Thomaidis, N.S. Application of an advanced and wide scope non-target screening workflow with LC-ESI-QTOF-MS and chemometrics for the classification of the Greek olive oil varieties. *Food Chem.* **2018**, *256*, 53–61. [CrossRef] [PubMed]

39. Guo, J.; Yue, T.; Yuan, Y.; Wang, Y. Chemometric classification of apple juices according to variety and geographical origin based on polyphenolic profiles. *J. Agric. Food Chem.* **2013**, *61*, 6949–6963. [CrossRef]

40. Kahle, K.; Kraus, M.; Richling, E. Polyphenol profiles of apple juices. *Mol. Nutr. Food Res.* **2005**, *49*, 797–806. [CrossRef]

41. Hyson, D.A. A comprehensive review of apples and apple components and their relationship to human health. *Adv. Nutr.* **2011**, *2*, 408–420. [CrossRef]

42. Sun, J.; Lin, L.Z.; Chen, P. Study of the mass spectrometric behaviors of anthocyanins in negative ionization mode and its applications for characterization of anthocyanins and non-anthocyanin polyphenols. *Rapid Commun. Mass Spectrom.* **2012**, *26*, 1123–1133. [CrossRef]

Article

Food Supply Chain Stakeholders' Perspectives on Sharing Information to Detect and Prevent Food Integrity Issues

Fien Minnens [1],*, Niels Lucas Luijckx [2] and Wim Verbeke [1]

[1] Department of Agricultural Economics, Faculty of Bioscience Engineering, Ghent University, Coupure Links 653, 9000 Ghent, Belgium; wim.verbeke@ugent.be

[2] TNO, Utrechtseweg 48, 3704 Zeist, The Netherlands; niels.lucasluijckx@tno.nl

* Correspondence: fien.minnens@ugent.be; Tel.: +32-9264 5925

Received: 27 May 2019; Accepted: 21 June 2019; Published: 25 June 2019

Abstract: One of the biggest challenges facing the food industry is assuring food integrity. Dealing with complex food integrity issues requires a multi-dimensional approach. Preventive actions and early reactive responses are key for the food supply chain. Information sharing could facilitate the detection and prevention of food integrity issues. This study investigates attitudes towards a food integrity information sharing system (FI-ISS) among stakeholders in the European food supply chain. Insights into stakeholders' interest in participating and their conditions for joining an FI-ISS are assessed. The stakeholder consultation consisted of three rounds. During the first round, a total of 143 food industry stakeholders—covering all major food sectors susceptible to food integrity issues—participated in an online quantitative survey between November 2017 and February 2018. The second round, an online qualitative feedback survey in which the findings were presented, received feedback from 61 stakeholders from the food industry, food safety authorities and the science community. Finally, 37 stakeholders discussed the results in further detail during an interactive workshop in May 2018. Three distinct groups of industry stakeholders were identified based on reported frequency of occurrence and likelihood of detecting food integrity issues. Food industry stakeholders strongly support the concept of an FI-ISS, with an attitude score of 4.49 (standard deviation (S.D.) = 0.57) on a 5-point scale, and their willingness to participate is accordingly high (81%). Consensus exists regarding the advantages an FI-ISS can yield towards detection and prevention. A stakeholder's perception of the advantages was identified as a predictor of their intention to join an FI-ISS, while their perception of the disadvantages and the perceived risk of food integrity issues were not. Medium-sized companies perceive the current detection of food integrity issues as less likely compared to smaller and large companies. Interestingly, medium-sized companies also have lower intentions to join an FI-ISS. Four key success factors for an FI-ISS are defined, more specifically with regards to (1) the actors to be involved in a system, (2) the information to be shared, (3) the third party to manage the FI-ISS and (4) the role of food safety authorities. Reactions diverged concerning the required level of transparency, the type of data that stakeholders might be willing to share in an FI-ISS and the role authorities can have within an FI-ISS.

Keywords: food integrity; transparency; food supply chain; information sharing; stakeholder; food fraud

---

1. Introduction

Integrity challenges along the food supply chain have received increasing attention by food safety authorities, industry and media in recent years. Adverse impacts of the adulteration of food products extend far beyond direct effects on the food industry itself. Indirect effects relate, amongst others, to the loss of public and consumer confidence in food production, the food industry, and food safety and quality [1,2]; international trade distortions and disputes [3]; and effects on food policies and consumer politics [4]. Following the BSE (Bovine Spongiform Encephalopathy) crisis at the end of the 1990s, Verbeke and Ward [5] showed that negative media coverage largely outweighs similar amounts of positive coverage aimed at reassuring the market and consumers. Therefore, eliminating the grounds for negative press, such as by following food integrity issues, emerges as a key attention point for the food supply chain.

A variety of measures are being developed and applied to safeguard food safety and quality, and to detect and prevent food integrity issues by different actors, both technical and organizational. Ellis, Muhamadali, Haughey, Elliott, and Goodacre [6] stressed that the ever-expanding portfolio of analytical methods, techniques and technologies in food control and future pervasive and predictive computation will together take on the role of a technology-based capable guardian for food systems. Simultaneously, more than ever before, experts recognize that food integrity is a challenge that requires a joint strategy and coordinated efforts involving all stakeholders, and that a strengthening of the collaboration between industry and governments is necessary [7]. The development of an integrated private–public strategy requires clearly defined roles for each participating stakeholder and clarity and shared agreement on the specific purpose [8].

The Elliott review following the 2013 horsemeat incident introduced eight pillars of food integrity: consumers first, zero tolerance, intelligence gathering, laboratory services, audits, government support, leadership, and crisis management [9]. The recommendations that are formulated for these eight pillars refer multiple times to the need for data, information and intelligence sharing between stakeholders: *"There needs to be a shared focus by Government and industry on intelligence gathering and sharing. The Government should work with the Food Standards Agency (to lead for the Government) and regulators to collect, analyze and distribute information and intelligence; and work with the industry to help it establish its own 'safe haven' to collect, collate, analyze and disseminate information and intelligence."* [9] (p. 7). Following the horsemeat incident, several actions were taken, and new initiatives were set up. For example, in the United Kingdom, the incident led to the establishment of the Food Industry Intelligence Network (FIIN), and on a European level it led to the Food Fraud Network (FFN), both aiming at the type of intelligence gathering that the Elliott review recommended [7].

Information and data that could be relevant to identify potential issues of integrity in food supply chains are often initially only available to industry experts operating at a specific level of the agro-food supply chain. Ideally, this information and data would be shared, integrated and analyzed, in order to help reveal issues faster and more accurately, and to help prevent them. Although the integration of food integrity data and information covering the whole food supply chain in one digital system seems futuristic, the digital data revolution and developments in artificial intelligence are transforming many economic sectors already, and the food sector is often named as one that might benefit substantially from a similar transition too [10,11]. Several studies have explored ways to implement such a holistic approach to identify, detect or prevent food integrity issues. For example, Bayesian network models entail the potential for predicting increased likelihoods of the occurrence of incidents [12,13]. The potential adoption of the Internet of Things (IoT) in the agro-food supply chain seems promising but several challenges and constraints were identified by Brewster et al. [14]. Decentralized traceability systems such as Blockchain structures are gaining interest and could provide solutions to these constraints [15,16].

While the development of analytical methods, technologies, systems, and infrastructures is gaining momentum, questions relating to stakeholder acceptance, intentions, and their willingness to adopt and participate remain largely unaddressed thus far. It is important to assess stakeholders' attitudes towards data and information sharing and how information sharing systems will be received, with the goal of detecting and preventing food integrity issues. The aim of this study was to investigate food supply chain stakeholders' attitudes and intentions towards a food integrity data and information sharing system (further referred to as an FI-ISS). The consultation of stakeholders focused on three objectives which were addressed in three consecutive rounds of data collection. Firstly, we aimed to analyze how food industry stakeholders receive the idea of an FI-ISS, which preconditions they consider important and what explains their intention to join an FI-ISS or not. The second objective of this study was to determine key success factors for the successful development and adoption of an FI-ISS. Lastly, the study explored in further detail the meaning of the defined key success factors, specific sensitivities facing the introduction of an FI-ISS and the origins of eventual contentious points.

2. Materials and Methods

This study combined exploratory and descriptive conclusive research methods and consisted of three consecutive rounds of data collection in which different stakeholders active in the European food supply chain were consulted. The first round (November 2017–February 2018) focused on food industry actors ($n = 143$), while, during the second (March–April 2018) ($n = 61$) and third rounds (May 2018) ($n = 37$), the target group was broadened beyond food industry stakeholders alone. The type of stakeholders that participated and their distribution across stakeholder groups are presented in Table 1.

Table 1. Distribution of participating stakeholders in the three rounds of the stakeholder study.

Type of Stakeholder	n	%
First round		
Food industry	143	
Large (>250 employees)	60	42.0
Medium-sized (<250 employees)	22	15.4
Small (<50 employees)	21	14.7
Micro (<10 employees)	8	5.6
Not known	32	22.4
Second round		
Total number of participants	61	
Food industry	30	49.2
Research	10	16.4
Service to the food industry	9	14.8
Food safety authority	5	8.2
Law enforcement	3	4.9
Consumer organization	2	3.3
Other (e.g., consultants)	2	3.3
Third round		
Total number of participants	37	
Research	19	51.3
Food safety authority	8	21.6
Food industry	4	10.8
Government	2	5.4
Other (e.g., consultants)	4	10.8

The first round consisted of an online survey using a quantitative questionnaire, with the aim of identifying food industry actors' attitudes towards, views on, intentions to join and preconditions for the uptake of an FI-ISS. The questionnaire was web-programmed in Qualtrics and covered industry stakeholders' perceptions about food integrity issues, their attitudes towards the concept of sharing information to detect and prevent food integrity issues, and specific attitudes and intentions towards an FI-ISS. In order to inform participants and ensure consistent framing of the potential FI-ISS, participants were exposed to a 150 second explanation video that explained the main characteristics of an FI-ISS [17].

After watching the video, 143 stakeholders started the survey, of which 111 continued until the end of the survey. All descriptive results are presented for each question using all valid answers, i.e., including all data for those participants that have answered that specific question. The participants are stakeholders with diverse relevant responsibilities in their organization, with 87% of them being employed in quality or R&D management. The rest of the participants (13%) had responsibilities such as purchase management, general management, sales management or consultancy. Their businesses or organizations are active on different levels of the food supply chain such as processing (68%), primary production (25%), retail (21%) and services to the food industry (10%) such as software or packaging. Two-thirds are active in international trade. The food commodities they are actively working with cover the food commodities most vulnerable to food integrity issues [18]. All of the 111 participants that completed the survey from start to finish reported dealing with one or more of the most vulnerable commodities, namely, organic food (39%), milk (35%), grains (31%), spices (28%), fish (23%), olive oil (21%), honey and syrups (21%), fruit juices (19%), coffee and tea (17%), wine (10%) and meat (9%). This could be due to the fact that the topic of the survey, being food integrity issues, appeals more to those working in more vulnerable sectors, which might have increased their interest and motivation to participate in the study. Therefore, when interpreting these results, it is important to take the background of participants and possible self-selection bias into account.

First, participants were asked to report the frequency of occurrence of food integrity issues and their likelihood of detection within their own organization. Both items were measured on a 6-point categorical scale. Additionally, the perceived risk of food integrity issues was measured using three items on a 5-point Likert scale. Second, participants' attitudes towards information sharing with the aim of tackling food integrity issues were assessed using three items (negative–positive, uninteresting–interesting, unimportant–important) on a 5-point bipolar interval scale. In a similar vein, perceived usefulness of information sharing was measured using three items (useless–useful; irrelevant–relevant; unnecessary–necessary). Three items on a 5-point Likert scale were used to measure how stakeholders perceive the risk of food integrity issues including "My company is very concerned about becoming a victim of food fraud", "Food integrity issues are a growing problem in our sector" and "Food integrity issues are one of the main risks our company faces". Third, participants were exposed to a series of potential advantages ($n = 8$) and disadvantages ($n = 6$) of information sharing; conditions for joining ($n = 16$) an FI-ISS; intention to join an FI-ISS ($n = 7$); third parties ($n = 9$) that could manage the FI-ISS; types of data ($n = 9$) that could be shared; and minimal output of the FI-ISS ($n = 7$), and were asked to indicate their degree of agreement using 5-point Likert scales. The selection of possible advantages, disadvantages, conditions for joining, third parties, the output of the system and types of data was based on literature review and expertise within the research team. For each set of items, participants were provided with the option to add additional items if they felt crucial items were missing. The construct scores for perceived advantages, perceived disadvantages, perceived risk and intention to join an FI-ISS were obtained by aggregating the items, leading to mean scores. Internal consistency was assessed with Cronbach's α values. Finally, characteristics of the organization were recorded such as type of activity, food commodities or product groups covered, geographical scope (regional, national, pan-EU, global), and size (micro, small, medium-sized, large).

The results of the first round were summarized and presented to the participants of the second round with summarizing text and bar charts. The stakeholders involved in the second round were thus exposed to the insights obtained from round 1 and invited to provide feedback and additional comments with the aim of identifying key success factors for an FI-ISS. The survey in the second round consisted of five sections. It assessed perception of current food integrity issues; the potential of information sharing; suitable trusted third parties; types of data to share; and initiatives for setting up an FI-ISS. A combination of closed-ended and open-ended questions was used, allowing the collection of both quantitative data and qualitative insight. As a graphical aid, a selection of the bar charts (as shown in the results section of this paper) were presented to the participants of round 2 (Figures 1 and 2).

The third round of data collection consisted of an interactive workshop with stakeholders and experts involved in the food supply chain. The results of the previous rounds were shared with the participants by means of a plenary presentation, after which these were discussed in four parallel working group sessions with a moderator and following a discussion guide. Each working group provided feedback and conclusions, which were discussed in plenary. This third-round interactive workshop envisaged providing a better understanding of the meaning of different views and perspectives on the future application of an FI-ISS.

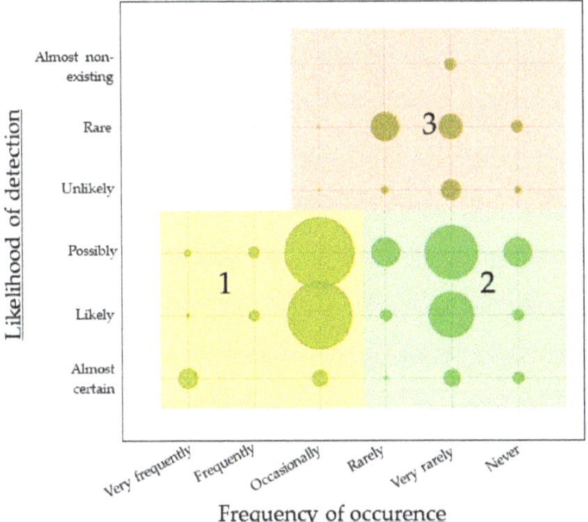

Figure 1. Bubble chart mapping food industry actors' perceptions of the occurrence of food integrity issues and the likelihood of detecting issues within their companies, identifying three clusters ($n = 133$, bubble sizes refer to n, 1 = Cluster 1, 2 = Cluster 2, 3 = Cluster 3).

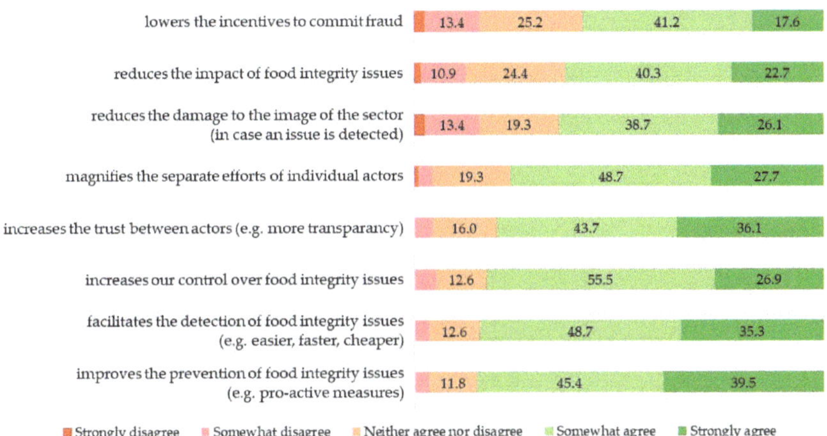

Figure 2. Perceived advantages of information sharing to prevent and detect food integrity issues according to food industry actors ($n = 119$, %).

In the interest of gathering unrestrained answers from food industry stakeholders, an anonymous approach was adopted for both data collection rounds involving online surveys. Before starting the survey, participants were informed about the context and purpose of the study, and researcher's contact details were provided. After completing the surveys, participants were invited to subscribe in a separate form (not linked to their responses) to receive feedback and an invitation for the next rounds. As such, the survey responses were never linked to personal identifiers and not linked between two rounds. All information was stored and processed in a non-identifiable format and reported in aggregated form only. Both surveys were distributed through an online link to a Qualtrics webpage. To reach a wide range of potential participants, multiple channels were contacted, and several federations and organizations agreed to share the survey links and workshop invitations within their professional networks, including the national federations of the Belgian, French and Dutch food industries (Fevia, ANIA, FNLI), specialized press such as Food Quality News and through the Food Integrity Network.

A challenge in this study was avoiding the dropout of participants during the consecutive rounds of the study. The questionnaire was concise and clear to limit the number of participants dropping out during the survey and to avoid survey fatigue. After the first round, it was explained to participants that they would receive the aggregate findings of the first round when staying involved in the study, with the aim of motivating them to take part in the next rounds. However, the possible efforts to avoid dropout were limited as a result of the anonymization, which made it impossible to send personalized reminders to participants. Statistical analyses were performed with SPSS Statistics 23.0 (IBM SPSS, Armonk, NY, USA), and qualitative responses to open-ended questions were analyzed with QSR International's NVivo 11 qualitative data analysis software (Melbourne, Australia). Cronbach's α values were computed to assess the internal consistency of measurement scales. Quantitative data processing and analysis included descriptive (frequency distributions), bivariate (correlations, chi-square tests, t-tests, and ANOVA) and regression analyses. Stepwise linear regression was adopted to identify predictors of the intention to join an FI-ISS. The Breusch–Pagan test was conducted to verify whether the assumption of homoscedasticity was satisfied. Qualitative data from open-ended questions were coded into categories for interpretation purposes.

3. Results and Discussion

3.1. Quantitative Survey with Food Industry Actors

3.1.1. Perceived Occurrence and Likelihood of Detection of Food Integrity Issues

A total of 143 industry stakeholders completed the questions about the occurrence of food integrity issues and the likelihood of the detection of issues within their own organization. The participants were categorized into clusters by combining these two variables. The chart in Figure 1 shows the frequency distribution of the reported occurrences and likelihoods of detection on the x- and y-axes, respectively, while the diameter of the circles represents the number of participants with that specific combination of responses. For example, an almost equal number of participants reported an occasional frequency of occurrence combined with the expected detection of the issue being classified as "likely" ($n = 16$) or "possibly" ($n = 17$).

Figure 1 also illustrates the three different clusters that were identified. Cluster 1 ($n = 51$) consists of food industry actors who perceive both the frequency of occurrence of issues and the likelihood of detecting issues as high. Cluster 2 ($n = 52$) is a group of actors who consider the frequency of food integrity issues to be low, and the likelihood that they would detect issues to be high. The smaller cluster 3 ($n = 30$) contains industry actors who perceive both the occurrence and likelihood of the detection of food integrity issues as low. Participants who indicated that they were unaware of the frequency of occurrence of food integrity issues within their organization ($n = 10$) were not included. The stakeholders in the three clusters differ significantly in terms of company sizes (Chi-square = 17.5; $p = 0.001$). Compared to the distribution in the sample, Cluster 3 is overrepresented by stakeholders from medium-sized enterprises (47.8%), while these are underrepresented in cluster 1 (5.1%). Cluster 2 consists of a balanced mix of stakeholders from different company sizes. This finding suggests that medium-sized companies perceive the likelihood of detecting food integrity issues as lower compared to smaller or larger companies. On one hand, smaller companies might have a better view and control over issues because of the smaller scale of the company. Larger companies, on the other hand, might feel more in control because of experience, economies of scales, or eventual measures they may have already taken to detect food integrity issues. There are no significant differences between the clusters in terms of business characteristics or attitudes towards information sharing (all $p > 0.05$). However, there is a difference in the perceived risk of food integrity issues. The mean perceived risk score (Cronbach's $\alpha = 0.67$) is 3.44 (standard deviation (S.D.) = 0.89) for the total sample. Stakeholders in cluster 1 perceive the risk as significantly higher ($\mu = 3.70$, S.D. = 0.90) compared to those in cluster 2 ($\mu = 3.26$, S.D. = 0.87) ($p = 0.045$).

3.1.2. General Attitude and Perceived Advantages and Disadvantages of Information Sharing

Food industry stakeholders' attitudes towards information sharing with the aim of tackling food integrity issues were measured on 5-point bipolar interval scales. The mean attitude score across the three items (Cronbach's $\alpha = 0.85$) was 4.49 ($n = 119$, S.D. = 0.57), which indicates that, in general, participants have a very positive attitude towards information sharing to detect and prevent food integrity issues. To measure participants' perception of the usefulness of information sharing, they were asked to rate three items on a 5-point bipolar interval scale. Cronbach's α for the three items was 0.81, indicating very good internal consistency reliability. The three item scores were aggregated and averaged to obtain an overall attitude score. The mean attitude score was 4.52 ($n = 119$, S.D. = 0.54), which indicates that, in general, participants perceive information sharing as very useful to detect and prevent food integrity issues.

In competitive environments, information sharing has advantages and disadvantages, depending on the aim and context for the different actors in the food supply chain and for the sector as a whole. First, participants were asked to score eight statements regarding potential advantages of information sharing. Figure 2 shows a strong consensus among stakeholders about most of the potential advantages, with less than 5% of participants disagreeing. However, disagreement is higher (more than 10%) for 'reduces the damage to the image of the sector', 'reduces the impact of food integrity issues' and 'lowers incentives to commit fraud'. The lack of consensus on these advantages implies that they are not obvious for all stakeholders, which suggests that possible doubts about the impact on the sector will need to be addressed when developing an FI-ISS. Discussions with stakeholders in the following rounds showed that there is a concern about the response, e.g., overreaction when issues are detected in an FI-ISS. The fear of overreaction might explain these doubts about the advantages for the sector. Although it is difficult to predict the impact of the adoption of an FI-ISS on the image of the food sector, Charlebois and Haratifar [19] concluded that the introduction of a food traceability system in the organic dairy sector was perceived as valuable by dairy consumers and suggested that information sharing entailed the potential to increase the market share of dairy products. To measure stakeholders' overall perception of the advantages, a mean aggregated score was calculated of the eight items (Cronbach's $\alpha = 0.770$). The mean of stakeholders' perceived advantages is 3.96 (S.D. = 0.55).

Sharing information within a food supply chain also might entail drawbacks for the stakeholders involved. The feasibility of an FI-ISS will depend strongly on its potential to avoid or tackle these pitfalls. Mapping the possible pitfalls or doubts of stakeholders enables a better definition of the requirements for an FI-ISS. Six statements were used regarding potential disadvantages of information sharing. The overall perception of the disadvantages of joining an FI-ISS was measured as a mean aggregated score of the six items (Cronbach's $\alpha = 0.748$), which amounted to 3.25 (S.D. = 0.64). No significant differences in perceived disadvantages were found between different company sizes, geographical scopes or clusters. Stakeholders' perceptions of disadvantages of an FI-ISS are negatively correlated with how they perceive the advantages ($r = -0.403$, $p < 0.001$).

Figure 3 shows that there is less consensus about disadvantages of information sharing compared to the level of agreement on its advantages. One of the main perceived disadvantages is that information sharing might increase the workload of staff, with 58.0% of participants confirming that they consider this a potential disadvantage. Hence, during the development of an FI-ISS, user-friendliness and the avoidance of too much additional workload should be taken into account. In a U.S. study where stakeholders evaluated options for improving product traceability, it was emphasized that the end solution should create net positive value for consumers and all companies in the food supply chain [20].

A potential FI-ISS has to be able to convince people that the overall benefits for food businesses will offset the costs. Over half of the stakeholders also showed distrust towards other actors, with the concern that they might share the wrong information or misuse the FI-ISS. In a similar vein, about one third of participants agreed that information sharing could have a negative impact on their competitive position. Interestingly, another third disagreed with this statement, i.e., they do not expect disadvantages in terms of competitive positions. In recent years, several experts have expressed their belief that food integrity should be considered a pre-competitive issue in a similar vein as food safety, i.e., a non-negotiable condition for market access. The Elliot Review urged the food industry to include food crime prevention in companies' corporate social responsibility policy and acknowledged that it can deliver commercial benefits [9].

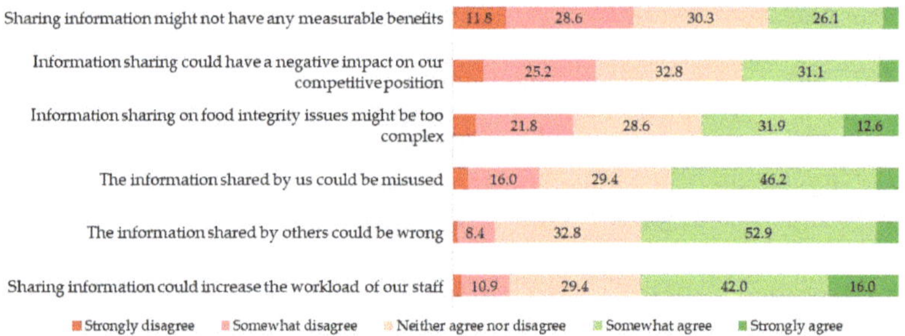

Figure 3. Perceived disadvantages of information sharing to prevent and detect food integrity issues according to food industry actors (*n* = 119, %).

3.1.3. Stakeholders' Conditions for Joining an FI-ISS

There is a clear consensus on several conditions that an FI-ISS needs to meet. First and foremost, industry actors agreed that they would need access to how the data and information that is shared will be handled, with the necessary protocols and procedures in place. Consequently, an important aspect is the confidentiality and encryption of the data and information, and also the anonymity of companies involved. The results show that industry actors consider this to be one of the main prerequisites; in other words, they would not join if their identity is not protected. There is also consensus on the importance of well-defined roles and rights of the trusted third party that manages the FI-ISS. Industry actors strongly agreed that data from authorities, research institutes, and NGOs should be incorporated in an information sharing system. Earlier research on the use of big data to identify emerging risks using Bayesian Network Modelling showed that several data sources can be used in predictive models [13]. For example, Bouzembrak et al. [21] showed how media reports on food fraud can be a useful input source for prediction systems. The participation of a sufficient number of actors in the sector in an FI-ISS is also a precondition for 82.3% of participants in our survey.

3.1.4. Intention and Its Determinants

Participants were asked about their likelihood to join an FI-ISS if their requirements were met. About 75% of participants reported that they would share all relevant information with an FI-ISS. Moreover, over 80% claimed that they would recommend an FI-ISS to their suppliers; 67% would recommend it to their customers, and 62% would recommend it to their competitors. The participants received a rather abstract description of the concept an FI-ISS, without any mentioning of funding or cost. However, in the survey, almost one third (31.5%) of participants expressed that they would be willing to pay for access to an FI-ISS.

The seven statements about the likelihood to join are considered a measure of the intention of stakeholders to join an FI-ISS. The aggregated score shows strong internal consistency (Cronbach's α = 0.879). Stakeholders' mean intention to join an FI-ISS is 3.81 (S.D. = 0.71). The intention to join an FI-ISS is significantly different for stakeholders from different company sizes. Medium-sized companies ($\mu = 3.47$, S.D. $= 0.71$) have a significantly lower intention to join an FI-ISS compared to large companies ($\mu = 3.97$, S.D. $= 0.68$). No significant differences in intentions were found in terms of other company characteristics such as geographical scope or cluster. On one hand, a positive and significant correlation was found between the intention to join an FI-ISS and the perceived advantages of an FI-ISS ($r = 0.280$, $p = 0.003$). On the other hand, a negative significant correlation was found between intention and perceived disadvantages ($r = -0.208$, $p = 0.028$). Although no significant correlation was found between a stakeholder's perception of the risk of food integrity issues and their intention to join, the perceived risk is positively correlated with the perception of advantages ($r = 0.293$, $p = 0.002$).

To assess the joint impact of these factors and company size as predictors of intention to join an FI-ISS, a stepwise linear regression analysis was adopted. With each step, explanatory variables are evaluated based on their significance level (Table 2). For the categorical variable size of the company, the category 'Micro and small' was considered the baseline category and 'Medium' and 'Large' were coded as dummy variables for comparison. The variables of perceived advantages, perceived disadvantages and perceived risk were entered as possible explanatory variables. The Breusch–Pagan test indicated that the assumption of homoscedasticity is satisfied ($p = 0.800$).

Table 2. Stepwise multiple regression: determinants of intention to join a food integrity information sharing system (FI-ISS) ($n = 111$).

Variables Entered	b	SE	β	t	p
Intention					
Constant	2.352	0.459		5.125	<0.001
Perceived advantages	0.393	0.115	0.304	3.412	0.001
Medium size	−0.480	0.158	−0.270	−3.028	0.003

b: unstandardized coefficient estimate; SE: standard error; β: standardized coefficient estimate; Model goodness-of-fit: Adjusted $R^2 = 13.5\%$.

Table 2 shows that perceived advantages and the size 'medium' were significant determinants of a stakeholder's intention to join an FI-ISS. Perceived advantages or benefits were also positively related to traceability systems implementation and the readiness for the implementation of a halal assurance system, for example, among small and medium-sized enterprises (SMEs) in the study of Abd Rahman et al. [22]. The dummy "medium size" is significant in the model with a negative coefficient, confirming that medium company size lowers the intention to join an FI-ISS. This may suggest different things, e.g., that medium-sized companies experience a lower urgency of dealing with food integrity issues, eventually as a result of having been less confronted with observed food integrity issues or simply being less convinced that information sharing might provide a solution to food integrity issues. This potential explanation is hypothetical, and this empirical finding is of interest to be further studied in the future. Other variables such as perceived disadvantage and perceived risk were not significant.

3.1.5. Suitability of Third Parties to Manage an FI-ISS

One of the main questions about organizing an FI-ISS is the choice of the trusted third party that will manage the system. A majority of 84.1% of participants agreed that a new organization established for this specific purpose would be the most suitable third party. This raises the question of whether the establishment of such a new organization can be realized in practice. Additionally, a food safety authority is also perceived to be a suitable third party by 74.3% of the industry stakeholders.

3.1.6. Information and Communication

An FI-ISS would gather information from different sources, analyze and manage it and produce an output of information which can be accessed by the members. For an FI-ISS to be effective, data and information need be shared by food businesses. There are three types of data which more than 70%

of participants agreed could be shared within an FI-ISS: monitoring and surveillance data, analytical data on product content, and data on certifications. However, it is important to note that for analytical data, 11.7% did not agree that these could be shared. It might be interesting to further analyze why a substantial share of industry stakeholders are reluctant to share that type of information. This coincides with 12.6% disagreeing with the potential sharing of information on the sourcing of their raw materials and ingredients. Less than half of the participating stakeholders confirmed that they would be willing to share import or trade data at company level, data on volumes or transactional data. The reluctance to share such commercially sensitive information can be due to the fear of losing competitive advantage. In addition, in a recent stakeholder study on the acceptance of blockchain technologies in meat supply chains, interviews with government officials showed that they doubted the feasibility of complete transparency and traceability [23]. The authors suggested the introduction of legal obligations to ensure that all actors provide complete information. In a similar vein, Bhatt et al. [20] pointed out that striking a balance between voluntary and regulatory compliance will be critical in the improvement of traceability.

An information sharing system can produce different types of output. The results of the survey show that stakeholders request clear protocols for action in case irregularities are detected. Additionally, stakeholders were very positive about the different possible outputs and agreed an FI-ISS should minimally produce a real-time searchable database, ad hoc alerts in case of irregularities and timely reports with filtered and analyzed information.

3.2. Insights from the Qualitative Survey

A broader range of stakeholders were involved in the second round of the study to gather different perspectives. The distinction was made between food industry actors, including participants that identified themselves as part of the food industry or services to the food industry, and other types of actors (including researchers, food safety authorities and law enforcement). All participating stakeholders were asked to analyze the response of industry actors obtained in round 1 and give their opinion on those results in a survey with both closed-ended and open-ended questions. The qualitative analysis of the answers from round 2 resulted in the selection of four key success factors (KSFs) for an FI-ISS. The four KSFs are presented in Figure 4, and each of the four will be discussed in detail in this section.

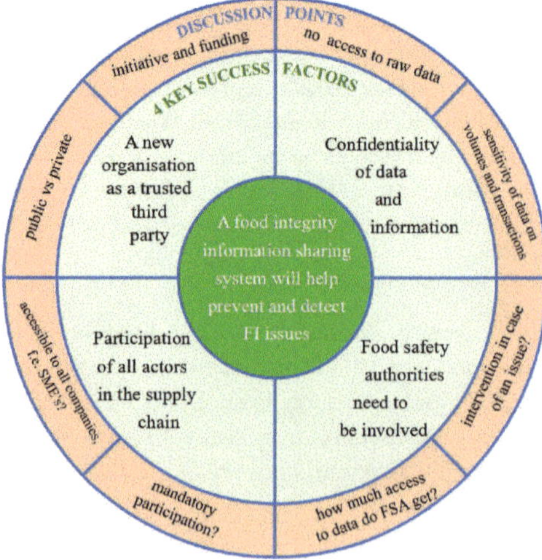

Figure 4. Four key success factors and associated discussion points for a food integrity information sharing system (FI-ISS). SME, small and medium-sized enterprise; FSA, food safety authorities.

3.2.1. Potential of an FI-ISS

The positive reception of the concept of an FI-ISS was presented to the participants in round 2. The participants in round 2 answered an open-ended question on why they think an FI-ISS has the potential to detect and prevent issues with food integrity, in addition to the advantages/disadvantages already covered in round 1. The answers were coded into categories. The advantages that were mentioned most are: information sharing will raise awareness on food integrity issues ($n = 12$), and the detection of food integrity issues will be easier and faster ($n = 11$). A number of stakeholders also mentioned that information sharing will give more insights into weak spots in the food supply chain ($n = 7$). However, some of the stakeholders mentioned risks or barriers that they perceived, inhibiting the potential of information sharing. For example, some referred to the risk that fraudsters might abuse the information in an FI-ISS ($n = 2$), the doubt that some indicators are not reliable enough because existing methods are not robust ($n = 2$) and the risk of potential overconfidence when having an FI-ISS in place ($n = 1$).

3.2.2. A New Organization as a Trusted Third Party

A first key success factor that became apparent from the results of the second round was the choice of a trusted third party. According to food industry actors in round 1, the two most suitable parties for organizing an FI-ISS were either a newly established organization or a food safety authority. Most of the stakeholders in round 2 confirmed that a newly established organization would be the most suitable trusted third party to manage an FI-ISS. Of the stakeholders ($n = 7$) who disagreed that a new organization should be the suitable third party, some mentioned the complexity of establishing and integrating another organization as the reason for their reluctance.

We also studied the idea of a new organization in more depth, by probing stakeholders' opinions about the criteria that such an organization should fulfill. Analysis of the open-ended questions shows that there are diverging opinions on the type of new organization that should be in charge of an FI-ISS. Some stakeholders believe that a new organization would need to operate on an international level ($n = 7$), arguing that food markets are globalized. The majority of stakeholders indicated to prefer a new organization established at the EU level ($n = 18$); however, they rarely clarified the reason for this preference. A total of 21 participants stated that they prefer a non-profit organization. One of the stakeholders who preferred a private organization mentioned *"confidential information will be involved"* as the reason for this preference. Another stakeholder used the same argument for his/her preference for a public organization, stating *"usually these food integrity issues involve sensitive data covered by data protection legislation"*. Several stakeholders mentioned why they think public funding should be used to finance an FI-ISS, for example *"if there is no charge to the companies using it then it would be more effective—cost might be an issue to SMEs where finances are tight"*. Another argument related to the primary purpose of an FI-ISS, where a participant mentioned *"it must be public since its primary purpose must be to protect consumers"*. It is important to note also that four stakeholders expressed a preference for a public–private partnership, without further argumentation as to why.

3.2.3. Confidentiality of Data and Information

Results from the first round showed there was a reluctance to share data on traded volumes and transactions. During the second round, stakeholders were asked for possible reasons for this reluctance. Analysis of the open-ended questions shows two main reasons. Firstly, most stakeholders mentioned the commercial sensitivity and the fear of competitive advantage issues. When setting up an information sharing system which aims to use the information on volumes and transactions, one needs to take these fears into account and foresee the necessary guarantees of anonymity and confidentiality of the information and data shared. Secondly, a few stakeholders also mention the fear of authorities having access to business information, more specifically finances and taxation authorities. Almost all stakeholders considered it necessary to anonymize data before they are seen by others.

These results confirm that there is no consensus on how much access to data actors in a system should really have. A considerable group considered it not possible to give all actors access to raw data.

3.2.4. Role of Food Safety Authorities (FSAs)

The majority of stakeholder participants in round 2 (78%) agreed that FSAs would be suitable as a trusted third party to manage an FI-ISS. FSAs exist on the national level but also on an international level, such as the EFSA (European Food Safety Authority). Stakeholders were asked for which of these they saw as having a role within an FI-ISS. The majority of participants indicated that this role can be for both (56%). Analysis of the comments made by stakeholders in the open-ended survey question which asked them to clarify their choice showed that eight stakeholders (consisting of one industry actor and seven non-industry actors) clearly expressed their preference for FSAs, considering them fit to manage an FI-ISS, and three reasons for this choice were expressed. Firstly, their already existing structure, network and expertise were mentioned, leading to a faster and less complex set-up of the system. Secondly, the already established trust among consumers was referred to. Thirdly, one stakeholder referred to the resources that FSAs have at their disposal. Another 12 stakeholders also mentioned that they see a role for FSAs and their expertise, but do not necessarily want them to be the third party that manages the FI-ISS, for example *"food safety authorities have an important role to play but are not equipped for this task. With the already existing systems they can play an important input role"*. Lastly, 13 stakeholders expressed the clear opinion that FSAs should not act as the third party managing an FI-ISS. A lack of trust among the industry was mentioned by four stakeholders, pointing at a lack of reliability and fear of sanctions. Furthermore, three stakeholders expressed their concern that FSAs should keep a certain distance from industry, for example, *"food safety authorities often have to maintain distance from industry, for fear of being assumed to be complicit with the food industry. Its action as a third party may prevent it from working appropriately"*. Another argument expressed by a few stakeholders is the difference between food safety issues and food integrity issues, and their worry that FSAs are not suitable to deal with the economic aspects of food integrity.

3.2.5. Participation of All Actors in the Food Supply Chain

The results from the first round showed that food industry actors find the participation of all actors a key success factor for an FI-ISS. In the second round, the survey probed for participants' opinions about the participation of consumer organizations and retailers. A number of stakeholders ($n = 10$) stated that they do not see a role for consumer or retail organizations, mentioning a few key arguments. Mostly, the commercial interest of retailers was an issue of concern. Stakeholders also mentioned lack of experience in dealing with food integrity issues, bad reputation and the fear that they would overreact to issues, for both retailers and consumer organizations. However, some stakeholders want a less active but rather advisory role for consumer organizations and retailers, for example, by being a member of the board of the new organization. Lastly, two stakeholders also considered their role in terms of communication and awareness raising, by sharing alerts, showing the efforts of the food industry and building consumer trust.

3.3. Insights from the Interactive Workshop

During the workshop, stakeholders were presented with an overview of the main results of the previous rounds. The four KSFs were presented, and stakeholders were invited to debate a number of results and questions during four parallel working group sessions, which were moderated using a common discussion guide. The discussion guide was developed and structured as indicated in Figure 4, covering the four possible KSFs for an FI-ISS and related discussion points to guide the discussion. Reflection with the four moderators after the sessions showed that the four different groups used the guide in various ways, from very directly following the questions to a looser approach where the stakeholders diverged from the suggested topics. Relevant statements that were made during the sessions are summarized in the discussion grid in Table 3.

Table 3. Summary of points raised by the stakeholders during the working group sessions in round 3 (May 2018).

Key Success Factors	Discussion Points
1. A new organization should manage an FI-ISS - Consensus in one of the groups about giving the task to a new organization - Some stakeholders prefer a new organization but not a new system, rather a system of systems, in a skeleton architecture or an umbrella - A technology company with the expertise and the know-how under a service contract could execute the technology component, with, on top of that, a public–private partnership for governance	**Should a new organization be private or public?** - Consensus in one of the groups that coordination and communication must be done by a public organization at EU level—funded by EU Commission—but a technical partner should be in charge of data management and data architecture - We expect higher consumer trust in a public organization compared to a private organization - Communication and collaboration between different (national, EU and global) organizations will be crucial - We need to take into account both consumer trust and industry trust, when setting up a system - Impartiality and conflict of interest are important when choosing a trusted third party **Who should take the initiative? Who should fund the system?** - The EU Commission should fund the initiative - One group found it unclear who should fund the new organization - Another group agrees that the initiative should be authority-driven
2. Data and information confidentiality needs to be guaranteed - Industrial actors can only be in favor of sharing data anonymously, as they are concerned for the possible economic loss - Industry actors are concerned about trust among the general population - Suggestion to share data on different levels, and adjust anonymity to the type of data (early indicators vs outcome indicators)	**Is it feasible to share in order to create a system, guaranteeing that nobody has access to the raw data that are shared?** - Focus should be on the sharing of meaningful data - Full anonymity is not possible: in case of an issue, the involved actors need to be able to be identified - There is a need to share metadata **Is the sensitivity of data on volumes and transactions a problem?** - Very low trust between industry actors, according to non-industry actors - Food integrity should be non-competitive, to try to take away some of the pressures of sharing data **Other** - At which point do authorities need to be informed of a food integrity issue? - Many data still on paper—need for a shift to machine readable data - By only sharing one-up and one-down, stove pipes are created. This is too restrictive
3. Food Safety Authorities (FSAs) need to be involved in the FI-ISS - Part of the issue is that FSAs lack the skillset, for example investigation skills	**Should FSAs have access to data and information?** - Currently it is a game of hide and seek between industry and authorities—industry is not willing to share all data with authorities - Authorities are bound legally **Uncertainty about reaction of FSAs in case of issues** - Worry confirmed: when will authorities be informed, and which information will they receive?
4. All actors in the supply chain need to be in the system - All European countries need to be involved - Inclusion of regulatory bodies, retail, NGOs - The question remains of whether actors with bad intentions will join?	**Should participation be mandatory?** - Ideally the system should be mandatory but stakeholders doubt that this is feasible - A mandatory system is not possible **Accessible for small and medium-sized enterprises (SMEs)?** - Small companies might be frightened to share data - Need for an incentive to convince all partners of the benefits of sharing data - Resource issue for very small companies—involve cooperatives or sector organizations to support SMEs

4. Conclusions

Dealing with complex food integrity issues requires a multi-dimensional approach. Besides the development of innovative analytical food process control methods, systems and practices, data and information sharing between actors in the food supply chain could help to facilitate the detection and prevention of food integrity issues. This study has demonstrated positive attitudes towards an FI-ISS among food industry stakeholders in the European food supply chain. Stakeholders are convinced of its benefits for detecting issues more rapidly and inexpensively, and for preventing the occurrence of issues. They believe that sharing information might magnify their own efforts, increase trust and improve the image of the food sector as a whole. Nonetheless, they are concerned about the increase in workload and the cost of such a system. An industry stakeholder's perception of the advantages of an FI-ISS is a predictor of their intention to join an FI-ISS. Though perceived disadvantages are negatively correlated with perceived advantages, this does not emerge as a significant determinant of intention to join. The perception of the risk of food integrity issues for their organization is not a significant predictor of their intention to join an FI-ISS. Importantly, food industry actors remain somewhat reluctant to trust the information that other actors might share, and they are doubtful about the reaction when a food integrity issue is detected. Medium-sized companies perceive the difficulties regarding the detection of food integrity issues as worse compared to small or larger companies, but still have a lower intention to join an FI-ISS. This empirical finding deserves further attention in future research.

Exposing a broader group of stakeholders to the results of the first round provided additional practical insights and led to the identification of four key success factors for an FI-ISS. Our study shows that the trusted third party that will manage an FI-ISS is preferably a new organization. However, a successful FI-ISS should also involve food safety authorities, albeit in an advisory role rather than a management one. As a third success factor, the study shows that the majority of stakeholders consider an FI-ISS to only be promising if data confidentiality is guaranteed by the data infrastructure. Our study showed that, in spite of their enthusiasm, most stakeholders are skeptical about the ways in which their information could be protected. The fourth key factor for the success of an FI-ISS is the participation of all actors in the food supply chain, including consumer organizations and retailers.

This stakeholder study faces limitations owing to its relatively small and self-selected sample of stakeholders. There may be selection bias as a result of higher involvement with food integrity issues among the study participants. Stakeholders' positive attitudes may, therefore, represent a best-case scenario, but are nevertheless encouraging for the development of an FI-ISS. The findings apply within the boundaries of the sample and generalization to the broader population of companies may be speculative. Insights on the barriers that might be encountered can be helpful for the food industry, food safety authorities and the science community in their efforts to ensure future food integrity, and eventually develop an effective FI-ISS.

Relatively high dropout levels of participants were encountered during the first round and between the consecutive rounds of the present study. Future studies might limit dropouts through stimulating the motivation and involvement of participants, e.g., by providing them with more detailed feedback and concrete benefits from the study findings. Another recommendation is to reduce the time lag between consecutive waves of data collection to maintain momentum and interest in the study topic.

As the development of an FI-ISS evolves, future socio-economic research could focus on monitoring the adoption process and collecting feedback about the experiences of food supply chain stakeholders. Eventually, cost-benefit analysis might be relevant to perform, and thus insight would be provided into what can be gained from the establishment and use of an FI-ISS. Additionally, the integration or linking of an FI-ISS to existing, private or public, traceability or alert systems requires further research.

Author Contributions: All authors contributed to the conceptualization and implementation of the research study; Writing—original draft preparation, F.M.; Writing—review and editing, N.L.L. and W.V.

Funding: The research leading to these results has been funded by the European Union's Seventh Framework Programme for research, technological development and demonstration under Grant No. 613688 (Food Integrity Project).

Acknowledgments: The authors greatly acknowledge all the stakeholders for their participation in the study. The authors highly appreciate and are thankful for the contributions of Isabelle Sioen from Ghent University, Department of Food Technology, Safety and Health, to the design of the study, the implementation of the stakeholder workshop, and earlier drafts of this manuscript.

Conflicts of Interest: The authors declare no conflict of interest.

References

1. Barnett, J.; Begen, F.; Howes, S.; Regan, A.; McConnon, A.; Marcu, A.; Rowntree, S.; Verbeke, W. Consumers' confidence, reflections and response strategies following the horsemeat incident. *Food Control* **2016**, *59*, 721–730. [CrossRef]

2. Kendall, H.; Kuznesof, S.; Dean, M.; Chan, M.Y.; Clark, B.; Home, R.; Stolz, H.; Zhong, Q.D.; Liu, C.H.; Brereton, P.; et al. Chinese consumer's attitudes, perceptions and behavioural responses towards food fraud. *Food Control* **2019**, *95*, 339–351. [CrossRef]

3. Cartin-Rojas, A. Food fraud and adulteration: A challenge for the foresight of veterinary services. *Rev. Sci. Tech. Off. Int. Epizoot.* **2017**, *36*, 1015–1024. [CrossRef] [PubMed]

4. Solomon, H. Unreliable eating: Patterns of food adulteration in urban India. *BioSocieties* **2015**, *10*, 177–193. [CrossRef]

5. Verbeke, W.; Ward, R.W. A fresh meat almost ideal demand system incorporating negative TV press and advertising impact. *Agric. Econ.* **2001**, *25*, 359–374. [CrossRef]

6. Ellis, D.I.; Muhamadali, H.; Haughey, S.A.; Elliott, C.T.; Goodacre, R. Point-and-shoot: Rapid quantitative detection methods for on-site food fraud analysis—Moving out of the laboratory and into the food supply chain. *Anal. Methods* **2015**, *7*, 9401–9414. [CrossRef]

7. Brooks, S.; Elliott, C.T.; Spence, M.; Walsh, C.; Dean, M. Four years post-horsegate: An update of measures and actions put in place following the horsemeat incident of 2013. *NPJ Sci. Food* **2017**, *1*, 5. [CrossRef]

8. Spink, J.; Moyer, D.C.; Whelan, P. The role of the public private partnership in food fraud prevention—Includes implementing the strategy. *Curr. Opin. Food Sci.* **2016**, *10*, 68–75. [CrossRef]

9. Elliott, C.T. *Elliott Review into the Integrity and Assurance of Food Supply Networks—Final Report—A National Food Crime Prevention Framework*; The National Archives: London, UK, 2014.

10. Fritsche, J. Recent developments and digital perspectives in food safety and authenticity. *J. Agric. Food Chem.* **2018**, *66*, 7562–7567. [CrossRef] [PubMed]

11. Rychlik, M.; Zappa, G.; Añorga, L.; Belc, N.; Castanheira, I.; Donard, O.F.X.; Kouřimská, L.; Ogrinc, N.; Ocké, M.C.; Presser, K.; et al. Ensuring food integrity by metrology and fair data principles. *Front. Chem.* **2018**, *6*. [CrossRef] [PubMed]

12. Bouzembrak, Y.; Camenzuli, L.; Janssen, E.; van der Fels-Klerx, H.J. Application of bayesian networks in the development of herbs and spices sampling monitoring system. *Food Control* **2018**, *83*, 38–44. [CrossRef]

13. Marvin, H.J.P.; Bouzembrak, Y.; Janssen, E.M.; van der Fels-Klerx, H.J.; van Asselt, E.D.; Kleter, G.A. A holistic approach to food safety risks: Food fraud as an example. *Food Res. Int.* **2016**, *89*, 463–470. [CrossRef] [PubMed]

14. Brewster, C.; Roussaki, I.; Kalatzis, N.; Doolin, K.; Ellis, K. Iot in agriculture: Designing a europe-wide large-scale pilot. *IEEE Commun. Mag.* **2017**, *55*, 26–33. [CrossRef]

15. Mao, D.H.; Wang, F.; Hao, Z.H.; Li, H.S. Credit evaluation system based on blockchain for multiple stakeholders in the food supply chain. *Int. J. Environ. Res. Public Health* **2018**, *15*, 21. [CrossRef] [PubMed]

16. Tian, F. A supply chain traceability system for food safety based on haccp, blockchain & internet of things. In Proceedings of the 2017 International Conference on Service Systems and Service Management, Dalian, China, 16–18 June 2017; pp. 1–6.

17. Minnens, F. A Food Integrity Information Sharing System. Available online: https://youtu.be/Akk9K6L_ECg (accessed on 26 May 2019).

18. European Parliament. European Parliament Resolution on the Food Crisis, Fraud in the Food Chain and the Control Thereof. Available online: http://www.europarl.europa.eu/sides/getDoc.do?pubRef=-//EP//NONSGML+TA+P7-TA-2014-0011+0+DOC+PDF+V0//EN (accessed on 26 May 2019).

19. Charlebois, S.; Haratifar, S. The perceived value of dairy product traceability in modern society: An exploratory study. *J. Dairy Sci.* **2015**, *98*, 3514–3525. [CrossRef] [PubMed]

20. Bhatt, T.; Buckley, G.; McEntire, J.C.; Lothian, P.; Sterling, B.; Hickey, C. Making traceability work across the entire food supply chain. *J. Food Sci.* **2013**, *78*, B21–B27. [CrossRef] [PubMed]

21. Bouzembrak, Y.; Steen, B.; Neslo, R.; Linge, J.; Mojtahed, V.; Marvin, H.J.P. Development of food fraud media monitoring system based on text mining. *Food Control* **2018**, *93*, 283–296. [CrossRef]

22. Abd Rahman, A.; Singhry, H.B.; Hanafiah, M.H.; Abdul, M. Influence of perceived benefits and traceability system on the readiness for halal assurance system implementation among food manufacturers. *Food Control* **2017**, *73*, 1318–1326. [CrossRef]

23. Sander, F.; Semeijn, J.; Mahr, D. The acceptance of blockchain technology in meat traceability and transparency. *Br. Food J.* **2018**, *120*, 2066–2079. [CrossRef]

Article

Assessment of Virgin Olive Oil Adulteration by a Rapid Luminescent Method

Raúl González-Domínguez [1,2,*], Ana Sayago [1,2], María Teresa Morales [3] and Ángeles Fernández-Recamales [1,2,*]

[1] Department of Chemistry, Faculty of Experimental Sciences, University of Huelva, 21007 Huelva, Spain
[2] International Campus of Excellence CeiA3, University of Huelva, 21007 Huelva, Spain
[3] Department of Analytical Chemistry, Faculty of Pharmacy, University of Seville, 41012 Seville, Spain
* Correspondence: raul.gonzalez@dqcm.uhu.es (R.G.-D.); recamale@dqcm.uhu.es (Á.F.-R.);
   Tel.: +34-959-219975 (R.G.-D.); +34-959-219958 (Á.F.-R.)

Received: 16 June 2019; Accepted: 23 July 2019; Published: 25 July 2019

Abstract: The adulteration of virgin olive oil with hazelnut oil is a common fraud in the food industry, which makes mandatory the development of accurate methods to guarantee the authenticity and traceability of virgin olive oil. In this work, we demonstrate the potential of a rapid luminescent method to characterize edible oils and to detect adulterations among them. A regression model based on five luminescent frequencies related to minor oil components was designed and validated, providing excellent performance for the detection of virgin olive oil adulteration.

Keywords: virgin olive oil; hazelnut oil; adulteration; luminescence

1. Introduction

Industrial processing of vegetable oils usually includes a refining step that removes almost all of the undesirable minor components, such as colored compounds, free fatty acids, or metals, but retaining the major neutral lipids and most of the natural antioxidants [1]. However, virgin olive oil is consumed crude, thus conserving the flavor compounds, vitamins, and other important natural components. Virgin olive oil adulteration is an important issue since the sensory and nutritional properties of this gourmet oil are responsible for its expensive price [2]. The most common adulteration process is the mixing of virgin olive oil with other cheaper oils, such as refined olive oil, seed oils (e.g., sunflower, soybean, corn, canola), as well as walnut, peanut, and hazelnut oils. This fraud is not easily detectable since refined oils lack of many of the minor compounds usually employed to authenticate olive oil, and the sensory characteristics of virgin oils cannot be significantly altered by this blending [3]. Particularly, hazelnut oil is commonly used for olive oil adulteration since its detection in mixtures is very challenging due to the very similar chemical profiles of these two oils [4]. This adulteration has been estimated to cause a loss of four million Euros per year for countries in the European Union. Therefore, there is a great interest in the development of accurate analytical methods, in combination with powerful chemometric tools [5], to guarantee the authenticity and traceability of virgin olive oil, and to detected olive oil adulteration. Frauds are usually detected by applying chromatographic techniques based on the determination of trans-fatty acids, sterols, triglycerides, hydrocarbons, and other components [6–8]. Other authors have also described the application of thermal analysis [9], and electroanalytical techniques [10–13], as potential approaches for investigating olive oil authenticity. Alternatively, spectroscopic techniques have also been proposed as suitable alternatives to traditional methodologies given their inherent advantages, including rapidity, environment-friendly nature, and low sample size requirement [14–16]. Several authors have described the application of vibrational spectroscopy (near-infrared spectroscopy, NIR; Fourier transform mid-infrared spectroscopy, FT-MIR; Fourier transform Raman spectroscopy, FT-Raman), NMR and visible absorption spectroscopy to investigate oil adulteration [17–19]. Among them, fluorescence spectroscopy enables rapid spectral acquisition with lower signal-to-noise ratio, so the combination of this technique with advanced chemometric tools has been successfully applied for the detection of olive oil adulteration in previous studies [20,21]. In this vein, the aim of this work was to evaluate the possibilities of a rapid luminescent method to detect fraudulent mixtures of virgin olive oils with refined hazelnut oils.

2. Materials and Methods

2.1. Oil Samples

Forty samples of four types of edible vegetable oils were directly acquired from oil mill stores from various locations: Eighteen virgin olive oil samples (VOO) of different European varieties (cvs. Cornicabra, Picual, Hojiblanca, Arbequina and Verdial were obtained from different producers in Spain; Cima di Bitonto and Tzunnati were of Italian and Greek origin, respectively; and a commercial sample (Borges brand, a blend of several olive oil varieties Tàrrega, Spain) were acquired in a market from Huelva in 2015), eight refined olive oils (ROO), seven virgin hazelnut oils (VHZO), and seven refined hazelnut oils (RHZO) (hazelnut oils were obtained from Turkish producers). Picual and Cima di Bitonto mixtures with refined hazelnut oils were prepared in the concentration range 5–30% (5, 10, 15, 20, 25 and 30% w/w) to evaluate the response of the luminescent spectra to the addition of adulterants. A test set was also prepared in the same concentration range using Cornicabra olive oil and refined hazelnut oil. Binary mixtures were prepared in triplicate. Oil samples were kept at 4 °C and analyzed without any pre-treatment to avoid potential interferences.

2.2. Instrumentation

All measurements were performed using a RF-1501 Shimadzu spectrofluorophotometer (Shimadzu Corporation, Kyoto, Japan) equipped with a continuous 150 W xenon lamp, and excitation and emission monochromators. Fluorescence emission spectra (300–800 nm, 1 nm interval) were collected at 650 nm excitation wavelength, while excitation and emission slits were set at 10 nm. Samples were scanned using a 3 mL non-fluorescent cell (10 mm path-length). After each series of measurements, the cuvette was cleaned using detergent, followed by a rinse with deionized hot water and acetone in order to dry and eliminate any remaining fat. Each sample was analyzed in triplicate. The spectrofluorophotometer was interfaced to a computer for spectral acquisition and data processing.

2.3. Data Analysis

Univariate and multivariate statistical analyses were conducted in Statistica 8.0 (Stat Soft, Tulsa, Oklahoma) and SIMCA-P™ software (version 11.5, UMetrics AB, Umeå, Sweden). Pattern recognition analysis was carried out by using unsupervised (principal component analysis, PCA; multidimensional scaling, MDS) and supervised (linear discriminant analysis, LDA; soft independent modeling of class analogy, SIMCA) chemometric techniques. One-way ANOVA was applied for the selection of wavelengths able to differentiate oil samples, and the response to the addition of adulterants was evaluated by stepwise linear regression analysis (SLRA). Before performing statistical analysis, data was submitted to different pretreatments and combination of pretreatments. Standard normal variate (SNV) transformation and first derivative were selected as the most suitable scaling procedures to remove undesirable factors in the spectral raw data and to correct for possible baseline shifts in the spectral data. First derivative was performed according to the Savitzky and Golay method with second-order smoothing polynomials through five points.

3. Results and Discussion

3.1. Spectral Characterization of Oil Samples

Figure 1 shows the characteristic luminescent spectra of extra virgin olive, refined olive, crude hazelnut, and refined hazelnut oil samples, which can be divided in four regions. Region A (300–400 nm) is mainly related to tocopherols and pigments and, as can be seen in Figure 1, allowed distinguishing virgin olive oils from the other edible oils studied in this work. Region B (400–500 nm) showed good correlation with conjugated dienes (K232), conjugated trienes (K270), and hydrolysis products [22], which are associated with oil quality. This might explain the clear discrimination between refined and non-refined oils in this luminescent region. Vitamin E is correlated with region C (500–600 nm) [22], thus making it possible to distinguish virgin olive oil from virgin hazelnut oil and refined oils. Finally, region D (600–800 nm) is mainly attributed to chlorophylls and pheophytines, pigments usually contained in virgin olive oil [23].

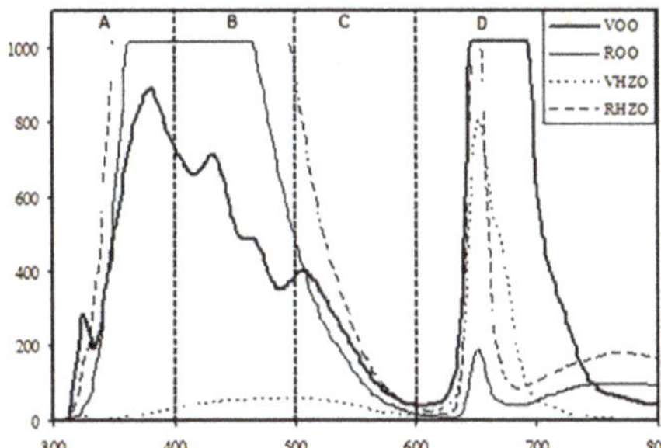

Figure 1. Characteristic luminescent spectra of virgin olive oil, virgin hazelnut oil, refined olive oil and refined hazelnut oil.

The method precision was estimated by computing the standard deviations (SD) and relative standard deviations (RSD) for each spectrum wavelength. However, results shown in Table 1 are presented as mean values for each of the regions from the spectrum. Repeatability was assessed by analyzing five replicates of a Cornicabra oil sample within the same day, while internal reproducibility was estimated by acquiring three replicate spectra of two different virgin olive oil samples in five different days by two different analysts. As shown in Table 1, repeatability and internal reproducibility were excellent for all the regions studied, with RSD values lower than 2% and 15%, respectively.

Table 1. Method repeatability (r) and internal reproducibility (R).

		Spectral Regions			
		300–400 nm	400–500 nm	500–600 nm	600–800 nm
r	SD	3.72	1.84	2.86	2.77
	RSD (%)	1.3	0.43	1.81	1.21
R	SD	14.96	22.27	17.33	6.07
	RSD (%)	10.69	6.01	7.63	3.34

3.2. Classification of Oils

As a first exploratory step, principal component analysis (PCA) and multidimensional scaling (MDS) were applied for a preliminary evaluation of data quality. PCA is an unsupervised method for reducing the dimensionality of the original data matrix retaining the maximum amount of variability, thus enabling us to get an overview of the data to identify possible outliers and trends towards the grouping of samples. On the other hand, the goal of MDS is to detect meaningful underlying dimensions that allow explaining observed similarities or dissimilarities between the investigated objects [24]. First, the acquired spectra were processed by using several spectral pretreatments, including first derivative, standard normal variate (SNV), and both first derivative and SNV. PCA was then performed on processed and normalized data, considering only factors with eigenvalues higher than 1 (Kaiser criterion). The best goodness of fit and validity was obtained by using spectral data in the region 650–800 nm ($R^2_X = 0.999$ and $Q^2_{cumulative} = 0.998$). This PCA model explained 98% of the total variance with four principal components (PC1 59.9%, PC2 28.3%, PC3 8.1%, and PC4 1.9), with PC1 being mainly associated with wavelengths in the ranges 656–720 and 752–800 nm. Figure 2A shows the distribution of samples in the plane defined by the first two principal components, which explained 88.2% of the original variance. Thus, it can be observed that samples were clustered in two groups, the first one comprising virgin olive oil samples, located in the right side of the plot, while the second cluster showed a high degree of overlapping between the other three groups of samples. The application of MDS also provided a clear separation of virgin olive oils from the rest of the sample set (Figure 2B). Dimension one discriminated between virgin olive oil samples and all refined samples as well as virgin hazelnut ones. Refined hazelnut and refined olive oils were clustered together, while virgin hazelnut samples were distributed in two groups, the first one near to the refined group, comprising three roasted samples, and the second one closer to the virgin olive oil cluster, constituted by three non-roasted crude hazelnut oil samples.

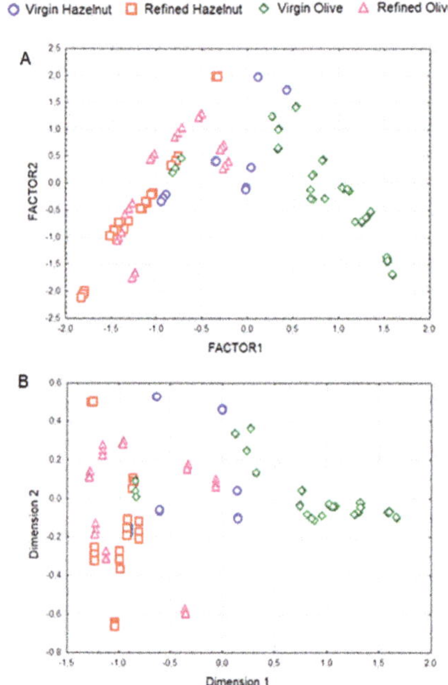

Figure 2. Principal component analysis (PCA) score plot (**A**) and multidimensional scaling plot (**B**) showing the distribution of samples from the four study groups in the plane defined by the two first principal components.

To achieve a more reliable differentiation among oil classes, various supervised pattern recognition procedures were also applied to the data matrix, including linear discriminant analysis (LDA) and soft independent modeling of class analogy (SIMCA). These classification methods have been successfully employed in previous studies to discriminate extra virgin olive oils according to variety and geographical origin [25,26], thus presenting a great potential in assessing food adulteration. LDA is a supervised classification tool based on the generation of orthogonal linear discriminant functions equal to the number of categories minus one. In this work, stepwise LDA was applied to classify oils according to the four categories, and the most significant variables involved in sample differentiation were selected using a Wilks' λ and F value as criterion for inclusion or removal of variables in the model. The best results were obtained when LDA was carried out on normalized data in the range 752–800 nm, with a mean prediction ability of 93%. The model retained seven variables (F to enter = 3.00 and F to remove = 1.00), showing a clear distinction among virgin oil samples and refined oil samples (Figure 3A). On the other hand, SIMCA is a class modeling technique that builds class models based on significant principal components of category, and classifies samples on the basis of their distance from the model representing each category. Models with five PCs were obtained by using SNV-pretreated data mean (99.7% of predictive ability), which explained 99.5, 99.9, 99.7, and 100% of the variance for virgin hazelnut, refined hazelnut, virgin olive, and refined olive categories, respectively. As shown in the corresponding Coomans plots, all vegetable oil samples could be correctly assigned (Figure 3B).

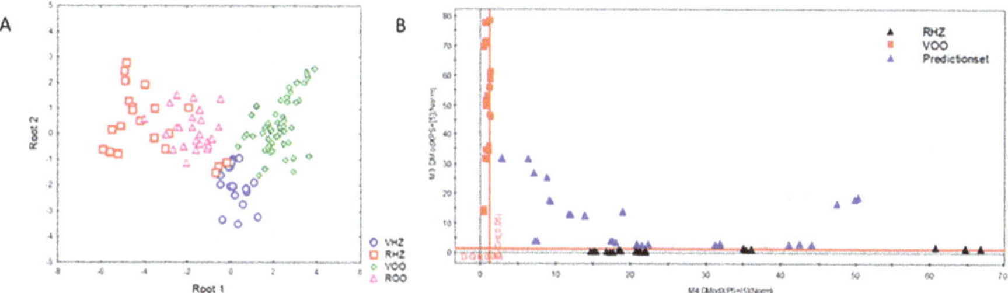

Figure 3. Linear discriminant analysis (LDA) score plot (**A**) and Coomans plot (**B**).

3.3. Detection of Virgin Olive Oil Adulteration by Regression Analysis

The mixture of virgin olive oil with refined hazelnut oil is a common fraud since this adulteration is not easily detectable by tasting or smelling. For this reason, we conducted an ANOVA test to look for variables enabling the differentiation between these two oils. Only wavelengths attributable to pigments (i.e., chlorophyll, pheophytin) and tocopherols presented significant *p*-values, in line with previous results [20].

Subsequently, regression models were built to assess the potential of luminescence to detect olive oil adulteration, for which several binary mixtures were prepared and analyzed. To this end, stepwise multiple regression analysis (SLRA) was applied to mixtures prepared by adding refined hazelnut oil in the concentration range 5–30%. This studied concentration range was selected on the basis of previous reports demonstrating the difficulty of detecting this adulteration at low concentration levels [4]. Two sets of blends were prepared by using Picual and Cima di Bitonto virgin olive oils, which were selected on the basis of ANOVA results to cover maximum (Picual) and minimum (Cima di Bitonto) values for pigments and tocopherols. SLRA was applied to these sets of mixtures under the strictest conditions (F to enter = 8.00 and tolerance = 0.01) to select the variables to be included in the model. Thus, five wavelengths were selected (319, 446, 476, 685, and 704 nm) to get an adjusted-R^2 of 0.972 (R^2 = 0.98; p = 0.000001; Durbin-Watson d statistic = 2.02). The linearity of the model (Figure 4) was excellent in the studied concentration range (5–30% w/w), with slope and intercept values close to 1 and 0, respectively. The statistical validity of this model was assessed by ANOVA through the lack of fit F tests. The ratio between the mean square due to the lack of fit and the mean square due to the pure experimental error was calculated (F = 2.45) and compared with the tabulated F value (F = 2.93), thus evidencing good adjustment between the observed and predicted values. Thus, the root mean square error of calibration (RMSEC) computed for this SLRA model was 0.77.

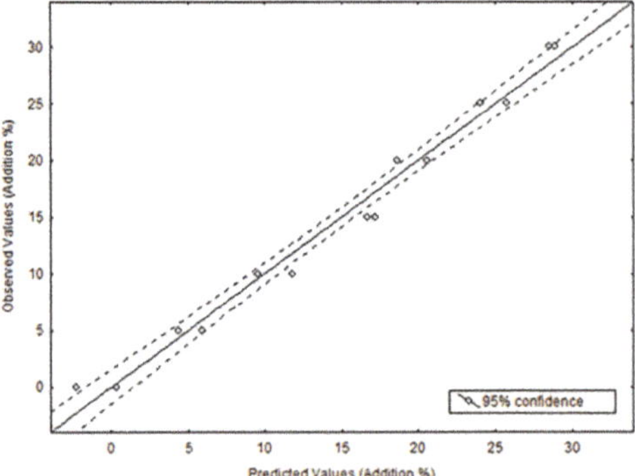

Figure 4. Prediction of adulteration percentage applying stepwise multiple linear regression to virgin olive oil and refined hazelnut oil mixtures.

Finally, a test sample was also prepared by mixing Cornicabra virgin olive oil with refined hazelnut oil in the same concentration range in order to evaluate the predictive ability of the model. The validation of the regression equation (Figure 4) in these test samples yielded adjusted-R^2 = 0.99 (R^2 = 0.99; p = 0.00079). Then, the root mean square error of prediction (RMSEP) was calculated as a measure of the accuracy of the model to predict the response. In the present work, the SLRA model yielded the RMSEP = 1.15, thus demonstrating the excellent accuracy of the luminescent method to detect adulterations. The limit of detection (LOD), calculated as three times the standard deviation of the intercept divided by the validation curve slope, was equal to 2.3%.

3.4. Method Performance: Comparison with Traditional Approaches

To evaluate the suitability and advantages of the luminescent method here presented to detect virgin olive oil adulteration with hazelnut oil, Table 2 shows the performance of previously published methods in this field. Due to the chemical similarities between olive and hazelnut oils chromatographic and mass spectrometric methods focused on the determination of specific oil components, such as filbertone [27], Maillard products [28], phytosterols [29], tocopherols [30], fatty acids [31], proteins [32], and phospholipids [33], usually provide low sensitivity to detect this adulteration. Furthermore, these methods are usually time consuming and require the application of complex extraction procedures, thus hindering their implementation in the food industry practice. Alternatively, several authors have also proposed the use of spectroscopic approaches, including nuclear magnetic resonance, infrared, and Raman spectroscopy [34–37], with increased potential for detecting relatively low adulteration levels (5–10%, w/w). Similar performance has been described for some genetic methods based on the application of polymerase chain reaction and subsequent capillary electrophoresis analysis (PCR-CE) [31], or high resolution melting (PCR-HRM) [38].

The luminescent method developed in this work yielded limits of detection around 2%, clearly surpassing the performance of most of the previously described methodologies. Furthermore, it should be also noted the greater simplicity, low analytical cost, and rapidity of analysis of this luminescent method compared with other conventional approaches, thus facilitating its implementation in routine laboratory analysis. Therefore, the present study demonstrates the possibilities of luminescent methods for the genuineness assessment of virgin olive oil.

Table 2. Performance of previously published methods to detect virgin olive oil adulteration with hazelnut oil.

Method	Lowest Level of Adulteration Detected (w/w)	Reference
Chromatographic methods		
LC/GC (filbertone)	20–25%	[27]
LC (Maillard products)	5%	[28]
GC (phytosterols)	>30%	[29]
LC (tocopherols)	3%	[30]
GC (fatty acids)	Non detectable	[31]
Mass spectrometric methods		
MALDI-TOF-MS (proteins)	20%	[32]
MALDI-TOF-MS (phospholipids)	1%	[33]
Spectroscopic methods		
^{1}H/^{31}P NMR	5%	[34]
2D NMR	6.27%	[35]
FTIR	25%	[36]
FT-Raman, FT-MIR	8%	[37]
Genetic methods		
PCR-CE	5%	[31]
PCR-HRM	10%	[38]

4. Conclusions

In this work, we have developed a rapid and simple luminescent method, in combination with advanced chemometric tools, to characterize and classify edible vegetable oils with good prediction ability. Furthermore, a regression model based on five luminescent frequencies related was validated for sensitive detection of virgin olive oil adulteration with hazelnut oil, a common fraud in food industry. The main strengths of this methodology rely on the simplicity, fast and low cost analysis compared with conventional approaches for adulteration detection. As a limitation, it should be mentioned the non-automatable nature of this technique, which could be improved in the future by the use of flow-through fluorescence cuvettes. Therefore, this work clearly demonstrates the possibilities of luminescent methods for the genuineness assessment of virgin olive oil, potentially implementable in food industry and regulatory agencies as a routine tool for adulteration detection.

Author Contributions: Conceptualization, Á.F.-R.; methodology, A.S., M.T.M. and Á.F.-R.; software, Á.F.-R.; validation, A.S. and Á.F.-R.; formal analysis, A.S. and Á.F.-R.; investigation, R.G.-D., A.S., M.T.M. and Á.F.-R.; resources, A.S. and Á.F.-R.; data curation, R.G.-D., A.S. and Á.F.-R.; writing—original draft preparation, Á.F.-R.; writing—review and editing, R.G.-D., A.S. and Á.F.-R.; visualization, R.G.-D., A.S. and Á.F.-R.; supervision, Á.F.-R.; project administration, Á.F.-R.

Funding: This research received no external funding.

Conflicts of Interest: The authors declare no conflicts of interest.

References

1. Alba, J. Elaboración de aceite de oliva virgen. In *El cultivo del olivo*; Barranco, D., Fernández-Escobar, R., Rallo, L., Eds.; Ediciones Mundi-Prensa: Madrid, Spain, 2001.

2. Aparicio, R. Authentication. In *Handbook of Olive Oil: Analysis and Properties*; Harwood, J.L., Aparicio, R., Eds.; Aspen Publishers: Gaithersburg, MD, USA, 2000.

3. Aparicio, R.; Morales, M.T.; Alonso, V. Authentication of European virgin olive oils by their chemical compounds, sensory attributes, and consumers' attitudes. *J. Agric. Food Chem.* **1997**, *45*, 1076–1083. [CrossRef]

4. Zabaras, D. Olive oil adulteration with hazelnut oil and analytical approaches for its detection. In *Olives and Olive Oil in Health and Disease Prevention*; Preedy, V.R., Watson, R.R., Eds.; Academic Press: London, UK, 2010; pp. 441–450.

5. Oliveri, P. Class-modeling in food analytical chemistry: Development, sampling, optimization and validation issues—A tutorial. *Anal. Chim. Acta* **2017**, *982*, 9–19. [CrossRef] [PubMed]

6. Fang, G.; Goh, J.Y.; Tay, M.; Lau, H.F.; Li, S.F.Y. Characterization of oils and fats by ^{1}H NMR and GC/MS fingerprinting: Classification, prediction and detection of adulteration. *Food Chem.* **2013**, *138*, 1461–1469. [CrossRef] [PubMed]

7. Jabeur, H.; Zribi, A.; Makni, J.; Rebai, A.; Abdelhedi, R.; Bouaziz, M. Detection of Chemlali extra-virgin olive oil adulteration mixed with soybean oil, corn oil, and sunflower oil by using GC and HPLC. *J. Agric. Food Chem.* **2014**, *62*, 4893–4904. [CrossRef] [PubMed]

8. Jabeur, H.; Zribi, A.; Bouaziz, M. Extra-virgin olive oil and cheap vegetable oils: Distinction and detection of adulteration as determined by GC and chemometrics. *Food Anal. Meth.* **2016**, *9*, 712–723. [CrossRef]

9. van Wetten, I.A.; van Herwaarden, A.W.; Splinter, R.; Boerrigter, R.; van Ruth, S.M. Detection of sunflower oil in extra virgin olive oil by fast differential scanning calorimetry. *Termochim. Acta* **2015**, *603*, 237–243. [CrossRef]

10. Apetrei, C.; Rodriguez-Mendez, M.L.; de Saja, J.A. Modified carbon electrodes for discrimination of vegetable oils. *Sens. Actuators B-Chem.* **2005**, *111–112*, 403–409. [CrossRef]

11. Apetrei, I.M.; Apetrei, C. Detection of olive oil adulteration using a voltammetric e-tongue. *Comput. Electron. Agric.* **2014**, *108*, 148–154. [CrossRef]

12. Apetrei, C.; Gutierez, F.; Rodriguez-Mendez, M.L.; de Saja, J.A. Novel method based on carbon paste electrodes for the evaluation of bitterness in extra virgin olive oils. *Sens. Actuat. B-Chem.* **2007**, *121*, 567–575. [CrossRef]

13. Tsopelas, F.; Konstantopoulos, D.; Tsantili-Kakoulidou, A. Voltammetric fingerprinting of oils and its combination with chemometrics for the detection of extra virgin olive oil adulteration. *Anal. Chim. Acta* **2018**, *1015*, 8–19. [CrossRef]

14. Wójcicki, K.; Khmelinskii, I.; Sikorski, M.; Caponio, F.; Paradiso, V.M.; Summo, C.; Paqualone, A.; Sikorska, E. Spectroscopic techniques and chemometrics in analysis of blends of extra virgin with refined and mild deodorized olive oils. *Eur. J. Lipid Sci. Technol.* **2015**, *117*, 92–102. [CrossRef]

15. Garrido-Delgado, R.; Muñoz-Pérez, M.E.; Arce, L. Detection of adulteration in extra virgin olive oils by using UV-IMS and chemometric analysis. *Food Control.* **2018**, *85*, 292–299. [CrossRef]

16. Dankowska, A.; Małecka, M. Application of synchronous fluorescence spectroscopy for determination of extra virgin olive oil adulteration. *Eur. J. Lipid Sci. Technol.* **2009**, *111*, 1233–1239. [CrossRef]

17. Nunes, C.A. Vibrational spectroscopy and chemometrics to assess authenticity, adulteration and intrinsic quality parameters of edible oils and fats. *Food Res. Int.* **2014**, *60*, 255–261. [CrossRef]

18. Sun, X.D.; Lin, W.Q.; Li, X.H.; Shen, Q.; Luo, H.Y. Detection and quantification of extra virgin olive oil adulteration with edible oils by FT-IR spectroscopy and chemometrics. *Anal. Methods* **2015**, *7*, 3939–3945. [CrossRef]

19. Jiménez-Sanchidrián, C.; Ruiz, J.R. Use of Raman spectroscopy for analyzing edible vegetable oils. *Appl. Spectrosc. Rev.* **2016**, *51*, 417–430. [CrossRef]

20. Sayago, A.; Morales, M.T.; Aparicio, R. Detection of hazelnut oil in virgin olive oil by a spectrofluriometric method. *Eur. Food Res. Tech.* **2004**, *218*, 480–483. [CrossRef]

21. Poulli, K.I.; Mousdis, G.A.; Georgiou, C.A. Rapid synchronous fluorescence method for virgin olive oil adulteration assessment. *Food Chem.* **2007**, *105*, 369–375. [CrossRef]

22. Kyriakidis, N.B.; Skarkalis, P. Fluorescence spectra measurement of olive oil and other vegetable oils. *J. Aoac. Int.* **2000**, *83*, 1435.

23. Mínguez-Mosquera, M.I.; Gandul-Rojas, B.; Gallardo-Guerrero, L. Rapid method of quantification of chlorophylls and carotenoids in virgin olive oil by high-performance liquid chromatography. *J. Agric. Food Chem.* **1992**, *40*, 60–63. [CrossRef]

24. Avramidou, E.V.; Doulis, A.G.; Petrakis, P.V. Chemometrical and molecular methods in olive oil analysis: A review. *J. Food Process. Preserv.* **2018**, *42*, e13770. [CrossRef]

25. Sayago, A.; González-Domínguez, R.; Urbano, J.; Fernández-Recamales, Á. Combination of vintage and new-fashioned analytical approaches for varietal and geographical traceability of olive oils. *LWT Food Sci. Technol.* **2019**, *111*, 99–104. [CrossRef]

26. Sayago, A.; González-Domínguez, R.; Beltrán, R.; Fernández-Recamales, Á. Combination of complementary data mining methods for geographical characterization of extra virgin olive oils based on mineral composition. *Food Chem.* **2018**, *261*, 42–50. [CrossRef] [PubMed]

27. Flores, G.; Ruiz del Castillo, M.L.; Herraiz, M.; Blanch, G.P. Study of the adulteration of olive oil with hazelnut oil by on-line coupled high performance liquid chromatographic and gas chromatographic analysis of filbertone. *Food Chem.* **2006**, *97*, 742–749. [CrossRef]

28. Zabaras, D.; Gordon, M.H. Detection of pressed hazelnut oil in virgin olive oil by analysis of polar components: Improvement and validation of the method. *Food Chem.* **2004**, *84*, 475–483. [CrossRef]

29. Cercaci, L.; Rodriguez-Estrada, M.T.; Lercker, G. Solid-phase extraction-thin-layer chromatography-gas chromatography method for the detection of hazelnut oil in olive oils by determination of esterified sterols. *J. Cromatogr. A.* **2003**, *985*, 211–220. [CrossRef]

30. Chen, H.; Angiulic, M.; Ferrari, C.; Tombari, E.; Salvetti, G.; Bramanti, E. Tocopherol speciation as first screening for the assessment of extra virgin olive oil quality by reversed-phase high-performance liquid chromatography/fluorescence detector. *Food Chem.* **2011**, *125*, 1423–1429. [CrossRef]

31. Uncu, A.T.; Uncu, A.O.; Frary, A.; Doganlar, S. Barcode DNA length polymorphisms vs fatty acid profiling for adulteration detection in olive oil. *Food Chem.* **2017**, *221*, 1026–1033. [CrossRef] [PubMed]

32. De Ceglie, C.; Calvano, C.D.; Zambonin, C.G. Determination of hidden hazelnut oil proteins in extra virgin olive oil by cold acetone precipitation followed by in-solution tryptic digestion and MALDI-TOF-MS analysis. *J. Agric. Food Chem.* **2014**, *62*, 9401–9409. [CrossRef]

33. Calvano, C.D.; Ceglie, C.D.; D'Accolti, L.; Zambonin, C.G. MALDI-TOF mass spectrometry detection of extra-virgin olive oil adulteration with hazelnut oil by analysis of phospholipids using an ionic liquid as matrix and extraction solvent. *Food Chem.* **2012**, *134*, 1192–1198. [CrossRef]

34. Vigli, G.; Philippidis, A.; Spyros, A.; Dais, P. Classification of edible oils by employing ^{31}P and ^{1}H NMR spectroscopy in combination with multivariate statistical analysis. A proposal for the detection of seed oil adulteration in virgin olive oils. *J. Agric. Food Chem.* **2003**, *51*, 5715–5722. [CrossRef] [PubMed]

35. Gouilleux, B.; Marchand, J.; Charrier, B.; Remaud, G.S.; Giraudeau, P. High-throughput authentication of edible oils with benchtop Ultrafast 2D NMR. *Food Chem.* **2018**, *244*, 153–158. [CrossRef] [PubMed]

36. Ozen, B.F.; Mauer, L.J. Detection of hazelnut oil adulteration using FT-IR spectroscopy. *J. Agric. Food Chem.* **2002**, *50*, 3898–3901. [CrossRef] [PubMed]

37. Baeten, V.; Fernández Pierna, J.A.; Dardenne, P.; Meurens, M.; García-González, D.L.; Aparicio-Ruiz, R. Detection of the presence of hazelnut oil in olive oil by FT-raman and FT-MIR spectroscopy. *J. Agric. Food Chem.* **2005**, *53*, 6201–6206. [CrossRef] [PubMed]

38. Vietina, M.; Agrimonti, C.; Marmiroli, N. Detection of plant oil DNA using high resolution melting (HRM) post PCR analysis: A tool for disclosure of olive oil adulteration. *Food Chem.* **2013**, *141*, 3820–3826. [CrossRef] [PubMed]

Article

Classification of Hen Eggs by HPLC-UV Fingerprinting and Chemometric Methods

Guillem Campmajó [1],*, Laura Cayero [1], Javier Saurina [1,2] and Oscar Núñez [1,2,3]

[1] Department of Chemical Engineering and Analytical Chemistry, University of Barcelona, Martí i Franquès 1-11, E08028 Barcelona, Spain

[2] Research Institute in Food Nutrition and Food Safety, University of Barcelona, Recinte Torribera, Av. Prat de la Riba 171, Edifici de Recerca (Gaudí), Santa Coloma de Gramenet, E08921 Barcelona, Spain

[3] Serra Húnter Fellow, Generalitat de Catalunya, Rambla de Catalunya 19-21, E08007 Barcelona, Spain

* Correspondence: campma03@gmail.com; Tel.: +34-93-403-3706

Received: 21 June 2019; Accepted: 29 July 2019; Published: 1 August 2019

Abstract: Hen eggs are classified into four groups according to their production method: Organic, free-range, barn, or caged. It is known that a fraudulent practice is the misrepresentation of a high-quality egg with a lower one. In this work, high-performance liquid chromatography with ultraviolet detection (HPLC-UV) fingerprints were proposed as a source of potential chemical descriptors to achieve the classification of hen eggs according to their labelled type. A reversed-phase separation was optimized to obtain discriminant enough chromatographic fingerprints, which were subsequently processed by means of principal component analysis (PCA) and partial least squares-discriminant analysis (PLS-DA). Particular trends were observed for organic and caged hen eggs by PCA and, as expected, these groupings were improved by PLS-DA. The applicability of the method to distinguish egg manufacturer and size was also studied by PLS-DA, observing variations in the HPLC-UV fingerprints in both cases. Moreover, the classification of higher class eggs, in front of any other with one lower, and hence cheaper, was studied by building paired PLS-DA models, reaching a classification rate of at least 82.6% (100% for organic vs. non-organic hen eggs) and demonstrating the suitability of the proposed method.

Keywords: HPLC-UV; fingerprinting; food classification; hen eggs; principal component analysis; partial least square-discriminant analysis

---

1. Introduction

In the last years, the interest of society in the food they purchase and consume has been raised. In this line, products with value-added due to specific particularities such as organic production, protected designation of origin (PDO), protected geographical indication (PGI), or those with fair-trade certification, are now receiving special attention. These labels not only ensure and guarantee food quality and traceability, but also mean an increment in its price in comparison with conventional products.

Hen eggs are among the most commonly eaten foods worldwide, as they have a high nutritional value, cheap costs, and are widely employed in international cuisines. They consist of two parts: The egg white, which mainly consists of 85% water and 10% proteins (ovalbumin being the most abundant one) approximately, and the egg yolk, which is composed of almost 22% lipids [1,2]. Moreover, their intake provides all the essential amino acids, many vitamins (vitamin A, riboflavin, choline, vitamin B_{12}, and vitamin B_9), and minerals (phosphorus, potassium, iron, and zinc).

In Europe, where almost 8 million tons of hen eggs were produced in 2017 [3], rules on their trade regarding production, hygiene, labelling, and marketing are laid down by the European Union (EU) [4–6]. Thereby, according to the European labelling eggs rule, each A quality category egg, which are those destined for human consumption, has to contain an identifier number code in its shell. Among other information that can be found on it, such as the country of origin (two-letter ISO - International Organization for Standardization- abbreviation code), the province, the municipality, and the producer establishment, the kind of hens and the breeding method employed are indicated by the first number digit:

Digit 0 is related to organic eggs (O), which means that they come from authorized and certified organic production farms. Thus, hens are fed with grown pasture and organic farming products, without employing transgenic substances nor antibiotics. The animals have a minimum space of 4 and 6 hens/m^2 outdoors and indoors, respectively. These are the most expensive eggs.

Digit 1 corresponds to free-range hen eggs (FR). In this case, their diet is mainly based on prepared cereal pellets, although grass can also be eaten. Antibiotics are mixed with food if needed. Moreover, similar space conditions to organic eggs are established.

Digit 2 indicates barn hen eggs (B). Hens do not have outdoor access, as they live in densely populated vessels and therefore, their diet consists of the prepared pellets and there is no entrance of natural light. Further, antibiotics are systematically provided with feed.

Digit 3 for eggs from caged hens (C), which are the cheapest ones. In these cages, hens can barely move (the minimum space allowed is of 12 hens/m^2) and there is no access to natural light either. Medical additives are provided with feed.

Due to the huge amount of produced eggs, two different frauds can be practiced. On one hand, in accordance with the European legislation, hen eggs have to reach the consumers within the 21 days of being laid [7], and their expiration date has to be fixed not more than 28 days after laying [6]. As there is no way to confirm whether those that are for sale are within the stipulated periods, some producers label them with erroneous dates, therefore giving a longer time before reaching their expiration date [8]. On the other hand, it is also difficult to distinguish hen eggs regarding their type. Although organic bodies may ensure the compliance of the established regulations, due to the high cost of the evaluation systems, some producers and distributors regulate themselves without adopting any national certification standard, leaving then the opportunity for food fraud [9].

The egg price increase from category 3 to 0 makes them susceptible to fraud, since a low category egg could be labelled as a superior one. Several methodologies have been previously developed in order to address egg authentication. For instance, profiling fatty acid composition by gas chromatography (GC), fitted with flame ionization detector (FID), in combination with chemometric techniques was proposed for the verification of organic against conventional eggs [1,10]. However, relatively time-consuming methodologies are usually required in order to determine the total lipid and fatty acid composition from the samples, also involving derivatization steps before GC separation. In another study, the carotenoid profile acquired by high-performance liquid chromatography and ultraviolet detection (HPLC-UV) was performed to classify both organic and conventional eggs [11]. Besides, in some cases, the authenticity of organic eggs and the assurance on their origin, was also approached by evaluating the level of several elements, including rare earth elements [12–14].

As can be seen, most of the methods described in the literature for egg authentication are based on targeted profiling approaches, which are focused on the specific determination of a given group of known selected chemicals. However, up to now, no specific biomarkers have been found in order to address hen eggs classification regarding their labelled class. Since many factors will affect the chemical composition of these products, non-targeted fingerprinting strategies that involve the determination of non-selective signals related to a range of potential discriminating compounds (i.e., spectrum or chromatogram), are promising approaches to address food authenticity issues [15–19]. As an example, a spectroscopic technique such as near infrared (NIR), in combination with principal component analysis (PCA), was proposed to achieve the classification of different type of eggs found in Chinese markets [20]. In the present work, HPLC-UV fingerprints recorded at 250 nm were proposed as a source of discriminant signals for hen eggs classification according to their production method by PCA and partial least squares-discriminant analysis (PLS-DA).

2. Materials and Methods

2.1. Chemicals and Standard Solutions

All the employed chemicals were of analytical grade. In the sample treatment, the acetonitrile and water (LC-MS Chromasolv® quality) used were purchased from Sigma-Aldrich (St. Louis, MO, USA). Methanol (UHPLC-gradient grade) was obtained from PanReac AppliChem (Barcelona, Spain) and formic acid (≥98%) from Sigma-Aldrich. For the mobile phase, water was purified using an Elix 3 coupled to a Milli-Q system from Millipore Corporation (Burlington, MA, USA) and filtered through a 0.22 µm nylon membrane integrated into the Milli-Q system.

2.2. Instrumentation

An Agilent 1100 Series HPLC instrument equipped with a quaternary pump (G1311A), a degasser (G1322A), an autosampler (G1329A), a diode array detector (G1315B), and a PC with the Agilent Chemstation software, all of them from Agilent Technologies (Waldbronn, Germany), was employed. HPLC-UV fingerprints were obtained by reversed-phase mode using a Kinetex C18 porous-shell column (100 mm × 4.6 mm I.D., 2.6 µm particle size) from Phenomenex (Torrance, CA, USA) at room temperature. Chromatographic separation was performed under gradient elution mode, using 0.1% (*v/v*) formic acid aqueous solution (solvent A) and methanol (solvent B) as mobile phase components, following the next elution program: 0–20 min, linear gradient from 15% to 95% solvent B; 20–30 min, isocratic elution at 95% solvent B; 30–30.1 min, back to initial conditions; and from 30.1–35 min, at 15% solvent B for column re-equilibration. The mobile phase flow rate was 0.4 mL/min, and the injection volume was 5 µL. The HPLC-UV fingerprints were registered at 250 nm.

2.3. Samples and Sample Treatment

Characterization and classification studies were carried out by analyzing 173 hen egg samples purchased from local markets (Barcelona, Spain). Table 1 classifies them according to their typology and manufacturer and defines their specified size as well as the number of samples.

Table 1. Description of the egg samples analyzed.

Egg Type	Manufacturer	Egg Size	Number of Samples
Organic hen eggs (O)	ViuBi	M/L	23
Free-range hen eggs (FR)	Vall de Mestral	-	23
	Ous Roig (Ebre)	-	23
	Ous Roig	L/XL	22
Barn hen eggs (B)	Liderou	M	24
	Eroski	L	24
	Ous Roig	L	11
Caged hen eggs (C)	Eroski	M	12
	Eroski	L	11

Sample extraction was performed following a previously described method [21], with some modifications. Briefly, 0.3 g of homogenized egg sample were weighed in an Eppendorf tube (Deltalab, Rubí, Spain), mixed with 1 mL of an acetonitrile:water 80:20 (*v/v*) solution by stirring in a Vortex (Stuart, Stone, United Kingdom) for 30 s, and then, centrifuged (Allegra™ 64R Centrifugue, Beckman Coulter, L'Hospitalet de Llobregat, Spain) for 10 min at 14,000 rpm and 4 °C. The supernatant extract was then filtered through 0.22 µm filter (Scharlab, Sentmenat, Spain) and stored at −18 °C in 2 mL glass injection vials until HPLC-UV analysis.

Moreover, a quality control (QC), which aimed to evaluate the repeatability of the method and the robustness of the chemometric results, was prepared by mixing 50 µL of each sample extract. A QC and a blank of acetonitrile were injected every 10 randomly injected samples.

2.4. Data Analysis

PCA and PLS-DA calculations were made by using SOLO 8.6 chemometric software (Eigenvector Research [22], Manson, WA, USA). Theoretical background of these methods in a detailed way is addressed elsewhere [23].

X-data matrices in both PCA and PLS-DA analysis consisted of the HPLC-UV chromatographic fingerprints obtained at 250 nm (absorbance intensities vs. retention time), whereas the Y-data matrix in PLS-DA defined each sample class. In order to improve the data quality, HPLC-UV chromatograms were smoothed, baseline-corrected, aligned, and autoscaled. Scatter plots of scores from principal components (PCs), in PCA, and latent variables (LVs), in PLS-DA, were used to study the distribution of samples, revealing patterns that could be correlated to their characteristics. In order to build both PCA and PLS-DA models, the first significant minimum point of the cross validation (CV) error from a Venetian blind approach was considered to be the most appropriate number of PCs or LVs, respectively.

3. Results and Discussion

3.1. HPLC-UV Chromatographic Separation

This work aimed to develop a HPLC-UV fingerprinting approach for the classification and discrimination of hen eggs according to their labelled typology. Thus, in order to obtain the richest chromatographic fingerprints, after a slightly modified simple liquid–liquid extraction procedure [19], the obtained extract of a B egg sample was employed for the optimization of the chromatographic separation by reversed-phase mode using 0.1% aqueous formic acid and methanol as mobile phase components. In a first separation consisting of a universal gradient, where methanol increased from 10% to 90% in 30 min, several compounds with different peak intensities were detected, although most of them elute close to the column dead volume. Thereby, different initial methanol percentages as well as the combination of gradient and isocratic steps were tested. As a compromise between the number of detected peaks, resolution, and analysis time, a final gradient consisting of an increase of methanol from 15% to 95% in 20 min followed by an isocratic step at 95% methanol for 10 min was selected. Figure 1 shows the obtained HPLC-UV chromatographic fingerprint registered at 250 nm for a B egg sample with the proposed gradient program.

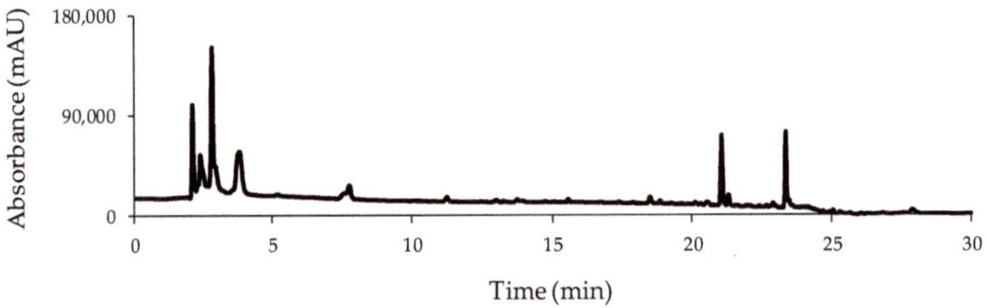

Figure 1. High performance liquid chromatography with ultraviolet detection (HPLC-UV) chromatogram at 250 nm obtained for a selected barn hen egg sample under the proposed gradient elution program (Section 2.2).

3.2. HPLC-UV Fingerprints

A total of 173 egg samples were analyzed by the proposed HPLC-UV method for classification purposes. For instance, Figure 2 shows the chromatograms at 250 nm for each one of the egg sample groups (O, FR, B, and C) analyzed. At a first glance, similar chromatographic fingerprints were obtained independently of the egg type. In fact, according to the retention times, the detected compounds seemed to be the same in each of them. However, variations associated to peak intensities, as well as their abundance within the different peak signals detected in a same sample, can be easily remarked. Therefore, these chromatographic fingerprints were evaluated and proposed as chemical descriptors to achieve sample classification.

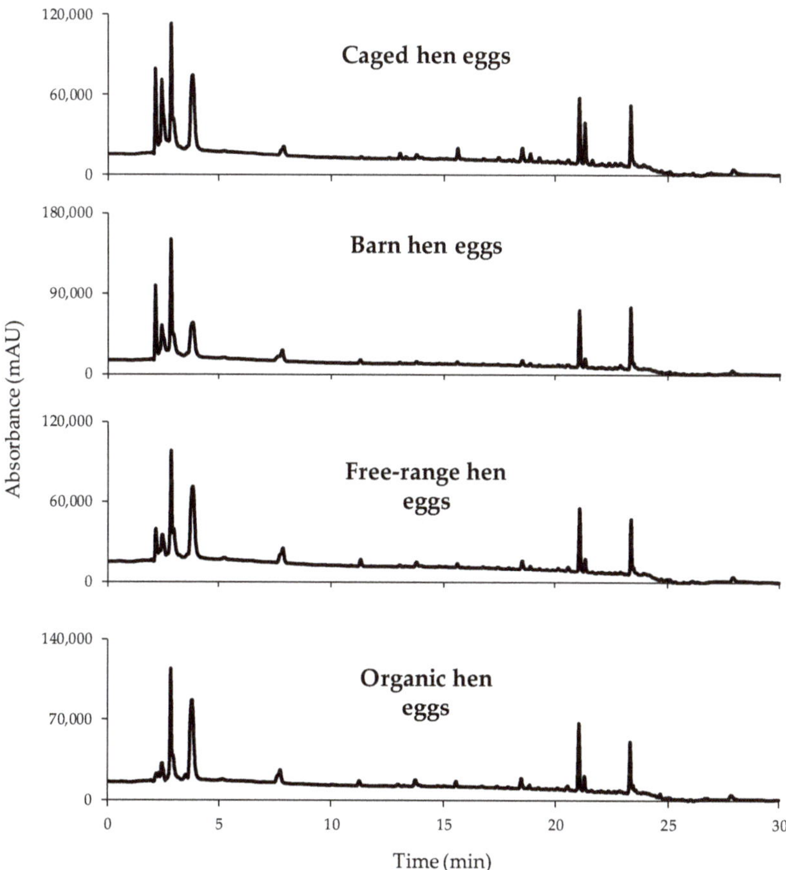

Figure 2. High performance liquid chromatography with ultraviolet detection (HPLC-UV) chromatographic fingerprints acquired at 250 nm for a selected sample within each egg type.

3.3. Classification of Samples According to Egg Type: PCA Study

As a first approach, a non-supervised PCA study was performed to evaluate the usefulness of HPLC-UV fingerprints for eggs classification according to their type. For that purpose, a data matrix (189 × 4506, samples × variables), which consisted of the recorded absorbance signals at 250 nm as a function of time for the analyzed egg samples and the QCs, was built. Moreover, data were pretreated as mentioned in Section 2.4, not only to reduce noise interferences, peak shifting, and baseline drifts, but also to provide the same weight to each variable by suppressing differences in their magnitude and amplitude scales. As a first result, the plot of scores of PC1 vs. PC2 (seven PCs were chosen for the PCA analysis), which is displayed in Figure 3A, shows that QC samples form a clear group (without any trend associated to a systematic error) in the upper side of the diagram, allowing the consideration of the obtained chemometric results. As can be seen in the plot of scores PC1 vs. PC3 shown in Figure 3B (seven PCs were also chosen for the supervised analysis), when excluding QC samples, even though there is not an evident discrimination among the samples, both the highest (O eggs) and lowest (C eggs) quality eggs predominate above and on the left of the plot, respectively. Up to this point, the proposed HPLC-UV fingerprints appeared to be adequate chemical descriptors at least for the distinction of O and C eggs, although PCA is only a non-supervised exploratory chemometric method. Therefore, in order to better exploit the obtained data and to improve the results on sample distribution, a supervised PLS-DA chemometric classification method was evaluated.

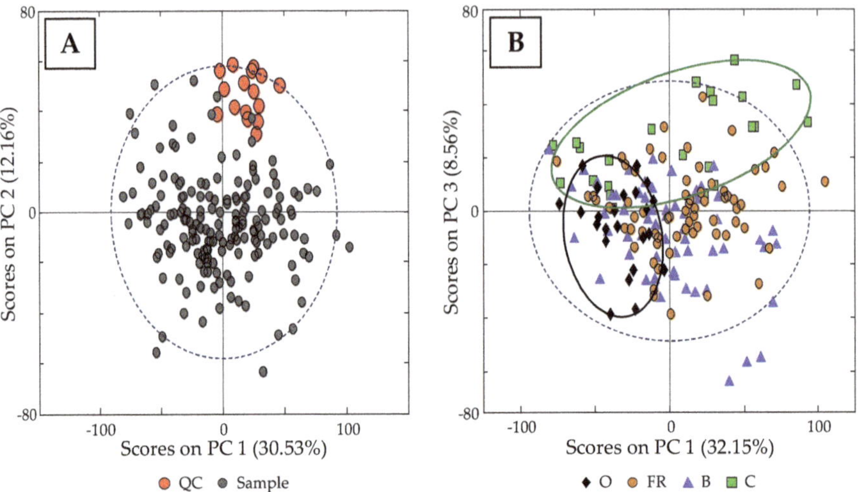

Figure 3. (**A**) Scores plot of PC1 vs. PC2 when using HPLC-UV fingerprints registered at 250 nm as chemical descriptors, showing a correct behavior of quality control (QC) samples. (**B**) Scores plot of PC1 vs. PC3, without including QC samples, showing a slight trend of organic hen (O) and caged hen (C) eggs.

3.4. Classification of Samples According to Egg Type: PLS-DA Study

The supervised chemometric study of all the analyzed egg samples was carried out by PLS-DA. For this reason, in addition to the X-data matrix previously described in the PCA study, a Y-matrix indicating the membership of each sample (O, FR, B, and C eggs) was used. The obtained scores plot of LV1 vs. LV2 (six LVs were chosen as optimum for the PLS-DA model, as detailed in Section 2.4), which is shown in Figure 4, improves non-supervised chemometric results as expected, and the obtained distribution seem to be directly related to the hens breeding method employed. In fact, eggs of hens with organic diet (O eggs) follow a particular trend mainly due to LV1, whereas LV2 affects those obtained from hens fed with a cereal-based diet and reared in cages (C eggs). On the other hand, in between these two group samples, both FR and B eggs, which as C eggs, are also collected from hens with a cereal-based diet but with better breeding conditions, apparently appeared randomly distributed.

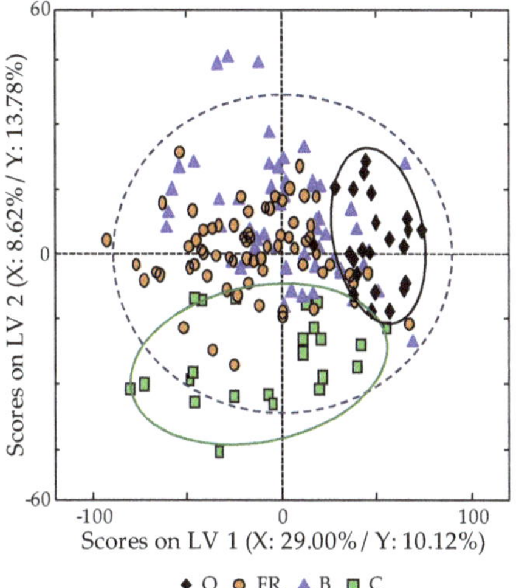

Figure 4. Partial least square-discriminant analysis (PLS-DA) scores plots of LV1 vs. LV2 when using HPLC-UV fingerprints registered at 250 nm as chemical descriptors.

A fact that should be taken into consideration is the number of manufacturers involved within the employed samples for each typology. While for O and C eggs all samples belonged only to one manufacturer, FR and B groups came from three. Thus, although according to the EU, same rules on the breeding process are established for a given quality egg class, additional sources of variance, such as the cereals employed in hens diet or the available grass and plants, could suppose a differential factor. Therefore, the applicability of HPLC-UV fingerprints as chemical descriptors to distinguish between the egg manufacturers was also evaluated by means of PLS-DA. For instance, Figure 5 shows the obtained scores plot of LV1 vs. LV2 when a 4 LVs PLS-DA model was built only for B egg samples. As can be observed, B eggs are clearly grouped according to their manufacturer, and thus, the proposed chromatographic fingerprints seem to be capable to remark these differences between different origins of production.

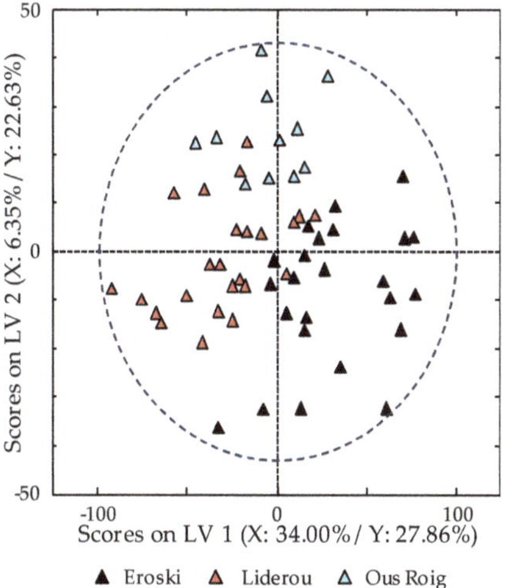

Figure 5. PLS-DA scores plots of LV1 vs. LV2 for B hen egg when using HPLC-UV chromatographic fingerprints registered at 250 nm as chemical descriptors.

Moreover, the size of the studied eggs was also evaluated by this methodology, as reported in Figure S1 (Supplementary Material). For that purpose, a matrix containing B and C egg samples, which were the only available classes labelled by size, was constructed. A clear distinction between M and L size eggs was achieved, independently of their class (B or C), denoting changes in the phytochemical fingerprint related to this morphological characteristic.

3.5. Supervised PLS-DA Method Validation

As the main goal of the present work was the discrimination of hen eggs according to their labelled class, and in order to demonstrate the applicability of the proposed method, the classification of higher class eggs in front of any other with one lower, and hence cheaper, was studied by building paired PLS-DA models (i.e., O vs. FR, B and C eggs; FR vs. B and C eggs; and B vs. C eggs). As can be observed in Figure S2 (Supplementary Material), the number of LVs employed to generate each classificatory model was selected considering the first significant minimum point of the CV error average as the most appropriate one.

For predicting the egg classes, the chemometric model was established using 70% of samples of each group as calibration set, while the remaining 30% was employed as "unknown" set for validation purposes. As can be seen in Figure 6A, O eggs, which are the most expensive ones, were clearly discriminated from those with lower prizes, reaching a classification rate of 100%. Further, while for the PLS-DA model of FR in front of B and C eggs (Figure 6B) a discrimination capacity of 82.6% was accomplished, B in front of C eggs (Figure 6C) resulted to be of 88%.

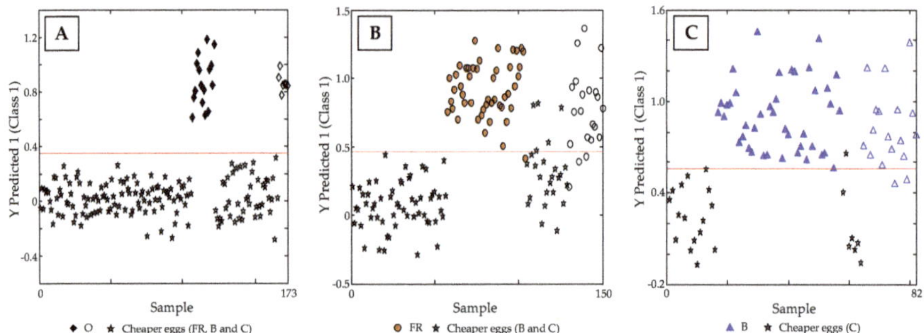

Figure 6. Sample vs. Y predicted 1 Scores plot for (**A**) O vs. FR, B and C eggs, (**B**) FR vs. B and C, and (**C**) B vs. C.

4. Conclusions

In this work, HPLC-UV chromatograms acquired at 250 nm have proved to be useful discriminant fingerprints for the classification of hen eggs according to their labelled typology. The PLS-DA models built for each egg category in front of those with lower one have reached at least a classification rate of 82.6%, showing satisfactory results of prediction. The distinction among organic and non-organic eggs has been especially satisfactory, in which 100% of sensitivity and selectivity has been reached. Moreover, the chromatographic fingerprints have also shown differences in egg phytochemical content among samples with different size independently of their type, as well as different manufacturers between samples from the same class.

Even though HPLC-UV fingerprints provided satisfactory results, the perfect classification of the four labelled hen egg groups was not achieved. At this point and in order to improve them, the evaluation of a new matrix, such as the egg yolk, rather than using the whole egg, could be an improvement to solve this problematic. Besides, fluorescence detection, which is characterized to be more selective than UV detection, could be proposed as an alternative detection technique for better descriptive models.

Finally, compared with biomarker-based strategies, the principal advantage of fingerprinting approaches is that the identification and quantification of selective species of each class are not essential for a successful sample classification. Here, in this regard, despite that specific markers have not been found, subtle differences in the content of components up- or down-expressed among classes have been exploited as the basis of the classification models. However, future work should also be directed towards biomarkers identification in order to address hen eggs authentication.

Supplementary Materials: The following are available online at http://www.mdpi.com/2304-8158/8/8/310/s1, Figure S1: PLS-DA scores plots of LV1 vs. LV2 for M and L size eggs when using HPLC-UV chromatographic fingerprints registered at 250 nm as chemical descriptors. Figure S2: Latent variable number vs. CV classification error average plots for the built PLS-DA models of: (A) O vs. FR, B and C eggs, (B) FR vs. B and C, and (C) B vs. C.

Author Contributions: G.C., J.S. and O.N. conceived and designed the experiments. G.C. and L.C. performed the experiments and processed the data. G.C. carried out the validation studies of all the chemometric methods proposed. G.C., J.S. and O.N. discussed the first results and suggested additional experiments. The first draft of the manuscript was prepared by G.C. and O.N. and it was revised and substantially improved by J.S.

Funding: This research was supported by the Agency for Administration of University and Research Grants (Generalitat de Catalunya, Spain) under the projects 2017SGR-171 and 2017SGR-310. G. Campmajó also thanks the University of Barcelona for the APIF fellowship.

Conflicts of Interest: The authors declare no conflict of interest.

References

1. Cherian, G.; Holsonbake, T.B.; Goeger, M.P. Fatty acid composition and egg components of specialty eggs. *Poult. Sci.* **2002**, *81*, 30–33. [CrossRef] [PubMed]
2. Fredriksson, S.; Elwinger, K.; Pickova, J. Fatty acid and carotenoid composition of egg yolk as an effect of microalgae addition to feed formula for laying hens. *Food Chem.* **2006**, *99*, 530–537. [CrossRef]
3. European Comission. Agriculture and Rural Development. Available online: https://www.ec.europa.eu/agriculture/eggs_en (accessed on 18 June 2019).
4. European Union. Council Directive 1999/74/EC of 19 July 1999 laying down minimum standards for the protection of laying hens. *Off. J. Eur. Communities* **1999**, *L203*, 53–57.
5. European Union. Commission Directive 2002/4/EC of 30 January 2002 on the registration of establishments keeping laying hens, covered by Council Directive 1999/74/EC. *Off. J. Eur. Communities* **2002**, *L30*, 44–46.
6. European Union. Commission Regulation (EC) No 589/2008 of 23 June 2008 laying down detailed rules for implementing Council Regulation (EC) No 1234/2007 as regards marketing standards for eggs. *Off. J. Eur. Union* **2008**, *L163*, 6–23.
7. The European Parliament and the Council of the European Union. Regulation (EC) No 853/2004 of the European Parliament and of the Council of 29 April 2004 laying down specific hygiene rules for food business operators on the hygiene of foodstuffs. *Off. J. Eur. Union* **2004**, *L139*, 151.
8. Johnson, A.E.; Sidwick, K.L.; Pirgozliev, V.R.; Edge, A.; Thompson, D.F. Metabonomic Profiling of Chicken Eggs during Storage Using High-Performance Liquid Chromatography-Quadrupole Time-of-Flight Mass Spectrometry. *Anal. Chem.* **2018**, *90*, 7489–7494. [CrossRef] [PubMed]
9. Rogers, K.M.; Van Ruth, S.; Alewijn, M.; Philips, A.; Rogers, P. Verification of Egg Farming Systems from the Netherlands and New Zealand Using Stable Isotopes. *J. Agric. Food Chem.* **2015**, *63*, 8372–8380. [CrossRef] [PubMed]
10. Tres, A.; O'Neill, R.; Van Ruth, S.M. Fingerprinting of fatty acid composition for the verification of the identity of organic eggs. *Lipid Technol.* **2011**, *23*, 40–42. [CrossRef]
11. Van Ruth, S.; Alewijn, M.; Rogers, K.; Newton-Smith, E.; Tena, N.; Bollen, M.; Koot, A. Authentication of organic and conventional eggs by carotenoid profiling. *Food Chem.* **2011**, *126*, 1299–1305. [CrossRef]
12. Barbosa, R.M.; Nacano, L.R.; Freitas, R.; Batista, B.L.; Barbosa, F. The Use of Decision Trees and Naïve Bayes Algorithms and Trace Element Patterns for Controlling the Authenticity of Free-Range-Pastured Hens' Eggs. *J. Food Sci.* **2014**, *79*, C1672–C1677. [CrossRef] [PubMed]
13. Borges, E.M.; Volmer, D.A.; Gallimberti, M.; Ferreira De Souza, D.; Luiz De Souza, E.; Barbosa, F. Evaluation of macro- and microelement levels for verifying the authenticity of organic eggs by using chemometric techniques. *Anal. Methods* **2015**, *7*, 2577–2584. [CrossRef]

14. Bandoniene, D.; Walkner, C.; Zettl, D.; Meisel, T. Rare Earth Element Labeling as a Tool for Assuring the Origin of Eggs and Poultry Products. *J. Agric. Food Chem.* **2018**, *66*, 11729–11738. [CrossRef] [PubMed]

15. Zhang, J.; Zhang, X.; Dediu, L.; Victor, C. Review of the current application of fingerprinting allowing detection of food adulteration and fraud in China. *Food Control* **2011**, *22*, 1126–1135. [CrossRef]

16. Riedl, J.; Esslinger, S.; Fauhl-Hassek, C. Review of validation and reporting of non-targeted fingerprinting approaches for food authentication. *Anal. Chim. Acta* **2015**, *885*, 17–32. [CrossRef] [PubMed]

17. Cuadros-Rodríguez, L.; Ruiz-Samblás, C.; Valverde-Som, L.; Pérez-Castaño, E.; González-Casado, A. Chromatographic fingerprinting: An innovative approach for food "identitation" and food authentication—A tutorial. *Anal. Chim. Acta* **2016**, *909*, 9–23. [CrossRef] [PubMed]

18. Sobolev, A.P.; Circi, S.; Capitani, D.; Ingallina, C.; Mannina, L. Molecular fingerprinting of food authenticity. *Curr. Opin. Food Sci.* **2017**, *16*, 59–66. [CrossRef]

19. Esteki, M.; Shahsavari, Z.; Simal-Gandara, J. Use of spectroscopic methods in combination with linear discriminant analysis for authentication of food products. *Food Control* **2018**, *91*, 100–112. [CrossRef]

20. Chen, H.; Tan, C.; Lin, Z. Non-destructive identification of native egg by near-infrared spectroscopy and data driven-based class-modeling. *Spectrochim. Acta Part A Mol. Biomol. Spectrosc.* **2019**, *206*, 484–490. [CrossRef] [PubMed]

21. Cavanna, D.; Catellani, D.; Dall'Asta, C.; Suman, M. Egg product freshness evaluation: A metabolomic approach. *J. Mass Spectrom.* **2018**, *53*, 849–861. [CrossRef] [PubMed]

22. Eigenvector Research Incorporated. Powerful Resources for Intelligent Data Analysis. Available online: http://www.eigenvector.com/software/solo.htm (accessed on 15 January 2019).

23. Massart, D.L.; Vandeginste, B.G.M.; Buydens, L.M.C.; de Jong, S.; Lewi, P.J.; Smeyers-Verbeke, J. *Handbook of Chemometrics and Qualimetrics*; Elsevier: Amsterdam, The Netherlands, 1997.

 *foods*

Article

Characterization and Differentiation of Spanish Vinegars from Jerez and Condado de Huelva Protected Designations of Origin

Enrique Durán-Guerrero [1,*], Mónica Schwarz [2], M. Ángeles Fernández-Recamales [3], Carmelo G. Barroso [1] and Remedios Castro [1]

[1] Analytical Chemistry Department—IVAGRO, Faculty of Sciences, University of Cadiz, 11510 Puerto Real, Spain
[2] "Salus Infirmorum" Faculty of Nursing, University of Cadiz, 11001 Cadiz, Spain
[3] Department of Chemistry, Faculty of Experimental Sciences, University of Huelva, 21007 Huelva, Spain
* Correspondence: enrique.duranguerrero@uca.es; Tel.: +34-956-01645

Received: 23 July 2019; Accepted: 9 August 2019; Published: 12 August 2019

Abstract: Thirty one Jerez vinegar samples and 33 Huelva vinegar samples were analyzed for polyphenolic and volatile compound content in order to characterize them and attempt to differentiate them. Sixteen polyphenolic compounds were quantified by means of ultraperformance liquid chromatography method with diode array detection (UPLC–DAD), and 37 volatile compounds were studied by means of stir bar sorptive extraction–gas chromatography–mass spectrometry (SBSE–GC–MS). Spectrophotometric CIELab parameters were also measured for all the samples. The results obtained from the statistical multivariate treatment of the data evidenced a clear difference between vinegars from the two geographical indications with regard to their polyphenolic content, with Jerez vinegars exhibiting a greater phenolic content. Differentiation by the volatile compound content was, however, not so evident. Nevertheless, a considerable differentiation between the two groups of vinegars based on their volatile fraction was achieved. This may bring to light the grape varieties and geographical factors that have a clear influence on such differences.

Keywords: vinegar; volatile compounds; polyphenolic compounds; differentiation

---

1. Introduction

Vinegar has been used since ancient times as a condiment and as a food preservative, along with many other uses [1]. Spain, and more specifically Andalusia, has an important wine-making heritage. Different vinegars are made using specific methods, thanks to the particular climate and ambience conditions in the region as well as the traditional manufacturing techniques that have been kept unaltered for many years. Some of the vinegar varieties that are produced in this region using such traditional and specific methods are nowadays recognized as protected designation of origin (PDO) products, a geographical indication that guarantees to consumers their quality and origin as well as a control system that governs the products from their production at the vineyard up to their commercialization [2]. Vinegar from Jerez and Condado de Huelva are two examples of such PDO products.

Jerez vinegar must invariably be obtained from the acetic fermentation of wines produced at the wine-making geographical indication known as "Jerez-Xérès-Sherry" and "Manzanilla Sanlúcar de Barrameda". The grape varieties that are allowed for vinegar-making are mainly Palomino and, to a lesser extent, Pedro Ximénez and Moscatel. These varieties are grown in the Jerez area and are adapted to its specific soil type, which is known as "albarizas". They are also adapted to the rainfall conditions and to the prevailing wind and temperature ranges in the area. All of these characteristics make this geographical indication a unique region with regard to cultivating conditions [3].

Vinegars that have been granted Condado de Huelva PDO are produced exclusively from wines that have been awarded the same PDO and are produced and aged within that specific environment. Zalema is the main grape variety that has been authorized for the production of this PDO vinegar. The varieties Palomino Fino, Listán de Huelva, Garrido Fino, Moscatel de Alejandría, and Pedro Ximénez are also used to a lesser extent. Similar to Jerez PDO, wine and vinegar from Condado de Huelva are also special, thanks to the unique environmental conditions and the compulsory specific oenological practices employed in their production [4].

In order to study and improve wine vinegar quality, it is essential to focus on its bouquet, an already widely studied feature that characterizes its particular olfactory and flavor profile [5–7]. Apart from its obvious culinary contributions, its beneficial properties for a healthy diet are increasingly valued. Natural antioxidant compounds deserve a special mention as their antiviral and anti-inflammatory properties as well as their capacity to assist the regulation of blood pressure [8] have already been studied in Jerez vinegars [9–11].

Some studies have examined the volatile compounds in vinegars to distinguish and classify them according to their raw material, production processes, and origin using different analytical methods coupled to gas chromatography (GC), such as headspace (HS) [12], solid-phase extraction (SPE) [2], solid-phase microextraction (SPME) [13,14], stir bar sorptive extraction (SBSE) [15,16], and headspace sorptive extraction (HSSE) [17]. Vinegars have also been differentiated by their polyphenol content by means of liquid chromatography [15,18] or spectroscopic techniques, such as infrared (IR) [19–21] or fluorescence [22], just to mention some of them.

Jerez and Huelva are two relatively nearby production areas (both of them are in the southwest of Andalusia in Spain), and both use similar procedures and the same raw material (grapes) for the production of vinegar. Nevertheless, geographic and climate-specific conditions may still play a relevant role in the final product. In fact, there are some studies where Andalusian vinegars from different production areas have been differentiated by their volatile [17] or polyphenolic content [18]. The grape varieties that are employed for different Andalusian vinegars under a PDO may be varied, but it could be possible to find vinegars from the same grape variety that exhibit some differences that are attributable to the geographical area in which they have been produced. Úbeda et al. detected some differences in the volatile content of red wines even though all of them had been produced from Carignan grapes but from different areas within the Chilean region known as Maule [23].

This study intends to characterize and differentiate different vinegar varieties from two geographical indications—Jerez and Condado de Huelva—based on their volatile and polyphenolic content by means of stir bar sorptive extraction–gas chromatography–mass spectrometry (SBSE–GC–MS) and ultraperformance liquid chromatography method with diode array detection (UPLC–DAD), respectively. This could be a useful tool to ensure the characteristic profile and authenticity of the products that are protected by their respective geographical indications.

2. Materials and Methods

2.1. Vinegar Samples

A total of 64 vinegar samples (33 Condado de Huelva PDO vinegar samples and 31 Jerez PDO vinegar samples) were analyzed in duplicate. In both cases, the samples were supplied by a representative number of wine cellars from each PDO.

2.2. Spectrophotometric L*a*b* Measurements

Color measurements were made using a Helios UV–VIS spectrophotometer (Unicam, Cambridge, United Kingdom) over the visible range 380–770 nm, with each interval being 10 nm. Color analyses were carried out following the method of the International Commission on Illumination (CIE) [24].

The CIELab coordinates L*, a*, and b* were obtained using the D65 illuminant and a 10° observer. In this three-dimension system, the L* axis indicates the lightness whose value extends from 0 (black) to 100 (white), while a* and b* axes represent the chromaticity. Coordinate a* has positive values for red colors and negative values for green colors. Coordinate b* has positive values for yellow colors and negative values for blue colors.

2.3. Determination of Polyphenols and Furanic Compounds

Polyphenol content was determined by UPLC according to the method improved by Schwarz et al. [25]. An Acquity UPLC equipment from Waters (Milford, MA, USA) coupled to a photodiode matrix ultraviolet–visible detector and an Acquity UPLC BEH C18 100 mm × 2.1 mm inner diameter column of 1.7 μm particle size (Waters, Milford, MA, USA). All the samples were filtered through 0.20 μm nylon filters (Millipore, Bedford, MA, USA) before their chromatographic analysis. In order to avoid possible losses of compounds during filtration, the amount of filtrated sample was large enough, and the initial volume was discharged.

The identification of each polyphenolic compound was carried out by comparing its retention time and ultraviolet–visible spectrum with those previously obtained from the analysis of standards. Calibration curves were established for each polyphenolic compound, with the concentration range taken into account according to bibliographic data.

The polyphenolic and furanic compounds that were quantified at 280 nm were as follows: gallic acid, 5-hydroximethyl-2-furaldehyde (HMF), furfural, tyrosol, syringic acid, protocatechuic acid, and protocatechuic aldehyde. The polyphenolic compounds quantified at 320 nm were as follows: caftaric acid, coutaric acid, caffeic acid, coumaric acid, syringaldehyde, ethyl caffeate, ethyl coumarate, and vanilline.

The polyphenolic compounds caftaric acid and ethyl caffeate were quantified by means of the caffeic acid calibration curve. Coutaric acid and ethyl coumarate were quantified by means of the coumaric acid calibration curve [25].

2.4. Determination of Volatile Compounds

Volatile compounds were determined by means of SBSE–GC–MS according to the conditions proposed by Durán et al. [26]. An aliquot of sample (25 mL) was pipetted for each SBSE analysis and poured into a 100 mL Erlenmeyer flask containing 5.85 g of NaCl and 84 μL of a solution of 4-methyl-2 pentanol (2.27 g/L in Milli-Q water containing 80 g/L of acetic acid). The Erlenmeyer flask was placed in a 15-position magnetic agitator (Gerstel, Mülheim an der Ruhr, Germany). Magnetic stirring bars that were 10 mm long and 0.5 mm wide and covered by an absorbing polymer called polydimethylsiloxane (PDMS) were used for the extractions. Known in the market as Twisters®, they were supplied by Gerstel (Mülheim an der Ruhr, Germany). The stirring bars were agitated at 1250 rpm for 120 min at 25 °C. After extracting the bars from the vinegar samples, they were rinsed with distilled water for a few seconds to remove the salt remains, and they were then pat-dried using cellulose paper. Then, they were inserted in a glass tube to be thermally desorped.

After that, the stirring bars were thermally desorped by means of a TDS-2 thermal desorption unit (Gerstel, Mülheim an der Ruhr, Germany) connected by a heat transfer line to a vaporizing injector at a programmed temperature (PTV) known in the market as Cis-4 (Gerstel, Mülheim an der Ruhr, Germany). The PTV was installed onto an Agilent 6890 GC-5973N MS (Agilent, Little Falls, DE, USA) system fitted with a capillary column DBWax (J&W Scientific, Folsom, CA, USA) of 60 m × 0.25 mm inner diameter and a 0.25 μm thick coating.

The compounds were identified by means of the Wiley library according to their analogy with mass spectra and confirmed by their standard retention indices or by the retention data found in the literature. Semiquantitative data of the compounds that had been identified were obtained by measuring the relative area of each compound's base peak (the most intense peak in the mass spectrum) in relation to the internal standard, 4-methyl-2-pentanol.

2.5. Statistical Analysis

The different data that were obtained were subjected to analysis of variance (ANOVA), principal component analysis (PCA), linear discriminant analysis (LDA), and cluster analysis (CA) by means of the computer application StatGraphics Centurion XVI (StatPoint Technologies, Inc., Warrenton, VA, USA).

3. Results and Discussion

3.1. Spectrophotometric L*a*b* measurements

L*a*b* parameters were measured for all the samples and ANOVA was performed. Table 1 shows the results obtained from the analysis.

Table 1. CIELab. Mean values and standard deviations of the parameters and analysis of variance.

	Jerez	Huelva	*F*-Ratio	*p*-Value
L* [+]	92.09 ± 3.24	89.44 ± 6.57	4.09	0.0476
a*	0.82 ± 1.17	1.73 ± 2.45	3.49	0.0664
b*	14.56 ± 7.54	19.46 ± 13.83	3.03	0.0866

[+] Significant differences for $p < 0.05$.

As can be seen, only lightness (L*) showed significant difference for $p < 0.05$, with Sherry vinegars presenting higher values for this parameter. Although a* and b* parameters did not present significant differences, vinegars from Huelva showed higher values, so they were more close to green and blue colors than Sherry vinegars.

3.2. Furanic and Polyphenolic Compounds

Table 2 shows the average values (mg/L) determined for the different polyphenolic and furanic compounds found. A statistical analysis of the data collected was carried out in order to verify if the origin of the vinegar samples would have any relevant influence on their actual polyphenolic content. The origin of the vinegar samples was taken as the independent factor. According to the results obtained for the "origin" factor, shown in Table 2, there were significant differences between the two groups of vinegars studied when $p < 0.01$. In that respect, the compounds that met this requirement were as follows: gallic acid, tyrosol, syringic acid, furoic acid, caftaric acid, coutaric acid, caffeic acid, coumaric acid, syringaldehyde, ethyl caffeate, ethyl cumarate, and vanilline (Table 2). Similar values were obtained by other authors when Jerez and Huelva vinegars were analyzed [10,18]. It can be seen that several of these compounds could be found with a significantly higher content in Jerez vinegars.

The resulting data were subjected to principal component analysis. This statistical tool searches for new variables to explain the maximum measured variability (polyphenols) between the different samples at the same time by taking into account the different correlations between them. The PCA is a statistical method that allows the number of variables to be reduced with the smallest possible amount of information getting lost in the process. The new principal components or factors are then expressed as a linear combination of the original variables and are also independent from each other. This allows the total variability to be explained while making use of a reduced number of factors.

According to Kaiser's criterion, three principal components were extracted to explain 78% of the variance between the samples. When Kaiser's criterion is applied, only components with eigenvalues greater than 1.00 are retained and interpreted. This is one of the most commonly used criterion and was proposed by Kaiser in 1960 [27]. The graphical representation of the samples on the orthogonal plane defined by the first two principal components can be seen in Figure 1. It can also be seen that PC1 was different depending on the geographical indication and that this PC took a positive value in the case of Jerez vinegars. The compounds with the greatest influence on PC1 were tyrosol, syringic acid, *p*-coumaric acid, syringaldehyde, and ethyl caffeate. The content of these polyphenols clearly increases when the vinegars are kept in wood containers, and this is something even more noticeable in Jerez vinegars [14].

Table 2. Polyphenolic and furanic compounds. Mean values of concentrations (mg/L) and analysis of variance.

Polyphenolic Compounds	Jerez (mg/L)	Huelva (mg/L)	*F*-Ratio	*p*-Value
Gallic acid *	50.7	36.7	7.78	0.0061
5-Hydroxymethyl-2-furaldehyde	70.9	66.9	0.04	0.84
Furfural	56	77.9	6.62	0.0112
Tyrosol *	20.8	15.6	14.25	0.0002
Syringic acid *	11.7	6.5	25.96	0
Protocatechuic acid	58.2	67.8	1.94	0.1661
Furoic acid *	64.5	172.6	34.36	0
Caftaric acid *	14.3	3.2	42.34	0
Coutaric acid *	13.5	3.4	73.59	0
Caffeic acid *	6.6	2.2	43.83	0
p-Coumaric acid *	8.1	1.7	71.36	0
Syringaldehyde *	7.3	3.6	50.84	0
Protocatechuic aldehyde	1.8	1.6	0.51	0.4785
Ethyl caffeate *	1.5	0.5	20.62	0
Ethyl coumarate *	1.2	0.1	20.62	0
Vanilline *	4.3	3	13.34	0.0004

* Significant differences for $p < 0.01$.

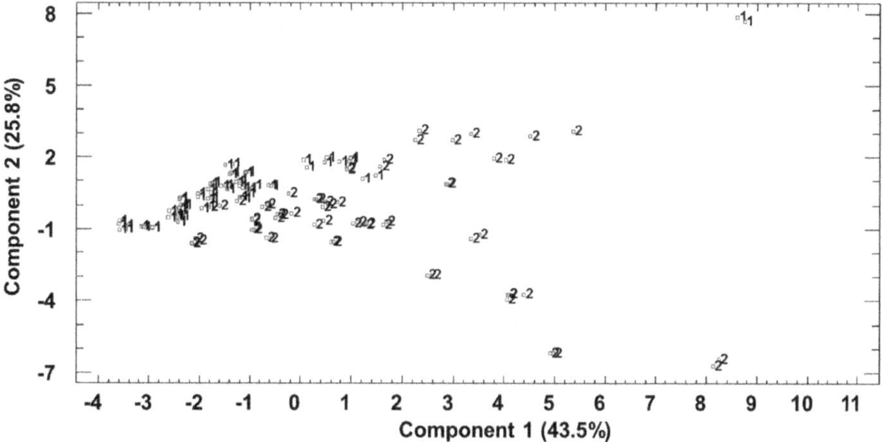

Figure 1. Principal component analysis of polyphenolic and furanic compounds. Projection of the samples onto the plane formed by the first two principal components. 1—Huelva; 2—Jerez.

A forward stepwise linear discriminant analysis with the leave-one-out cross-validation method was carried out later on. This type of analysis establishes which of the measured variables—in this case, polyphenolic compounds—contributes by a greater extent to a successful discrimination according to the geographical indication of each vinegar sample. Almost all the Huelva vinegar samples (96.88%) and 88.71% of the Jerez ones were successfully discriminated, meaning 92.86% of all samples were successfully classified. The compounds with the highest influence on the discriminant function were caftaric acid, *p*-coutaric acid, furfural, and syringaldehyde, among others.

Finally, the data on polyphenolic compounds were subjected to cluster analysis using Ward's method for clustering and the Euclidean square distance metric for comparison. This is a statistical multivariate technique that clusters elements (or variables) to obtain the greatest possible homogeneity within each group on the one hand and the greatest possible difference between groups on the other. The result is represented by a classification tree or dendogram in Figure 2.

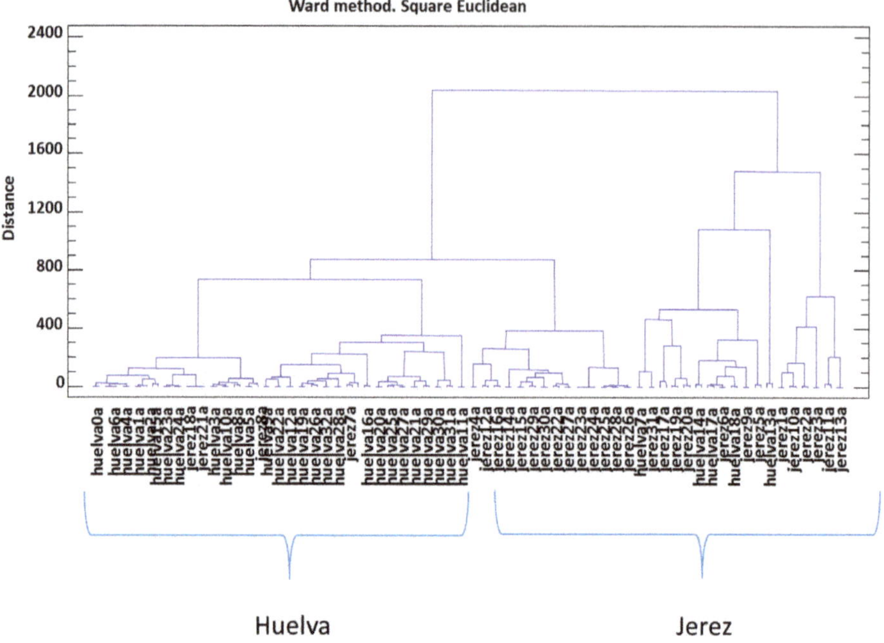

Figure 2. Cluster analysis of polyphenolic and furanic compounds.

It can be seen that two clear clusters were obtained, one with a large number of samples of Jerez vinegars and the other with Huelva vinegars. Nevertheless, some of the Jerez vinegar samples were classified together with Huelva vinegars. This fact may highlight how similar both groups are between them.

3.3. Volatile Compounds

Similar to the previous study on polyphenolic compounds, the analysis of variance based on a significant factor, i.e., the origin of the vinegar samples, was also carried out for the geographical indications Huelva and Jerez. Table 3 shows the results obtained from the ANOVA.

It can be observed that only a few of the volatile compounds demonstrated relevant differences in a single variance approach between the two geographical indications when the significance level was $p < 0.01$. These were 4-ethylguaiacol, 4-ethylphenol, 5-acetoxymethyl-2-furaldehyde, octanoic acid, decanoic acid, isobutanol ($p < 0.05$), isopentyl acetate, 3-hydroxy-2-butanone ($p < 0.05$), isovaleric acid, hexanoic acid ($p < 0.05$), and 2-phenylethanol. They exhibited a significant difference ($p < 0.01$) between vinegars from each of the two geographical indications, with generally higher value for vinegars from Jerez protected geographical indication (Table 3).

Subsequently, a PCA was carried out in order to cluster the differences in volatile content within a smaller number of variables (PC), which would explain the maximum variability between the samples. According to Kaiser's criterion (eigenvalue >1), 11 principal components were extracted, which explained 82% of the variance between the samples. The first two PCs accounted for 42.14% of the variability.

Table 3. Volatile compounds. Mean values of relative areas (RA) and analysis of variance.

Volatile Compounds	Jerez (RA)	Huelva (RA)	*F*-Ratio	*p*-Value
Eugenol	0.0121	0.0135	0.52	0.4735
Ethyl pentanoate	0.049	0.0047	67	0.4144
Ethyl lactate	1.9712	3.8071	0	0.9513
n-Butyl acetate	2.1283	1.1981	3.76	0.0566
trans-2-Hexen-1-ol	0.0074	0.0318	2.43	0.124
Furfural	0.0283	0.0282	2.15	0.147
Isobutyric acid	0.1154	0.4554	0.16	0.6948
2-Acetyl-5-methylfurfural	0.0171	0.013	0.02	0.8935
4-Ethylguaiacol *	0.1262	0.0141	48.24	0
4-Ethylphenol *	0.2041	0.0279	31.79	0
5-Acetoxymethyl-2-furaldehyde *	0.0311	0.0056	8.67	0.0044
Octanoic acid *	1.1339	0.6532	21.45	0
Decanoic acid *	1.7839	1.1319	12.4	0.0008
Ethyl isobutyrate	0.3935	0.4827	0.12	0.7349
Propyl acetate	0.0389	1.8364	1.8	0.1837
Isobutyl acetate	2.2364	4.7232	1.99	0.1626
Ethyl 2-methylbutyrate	0.1942	0.2033	0.87	0.3545
Ethyl isovalerate	1.853	1.7505	2.37	0.1281
Hexanal	0.0293	0.1588	0.64	0.4247
Isobutanol **	0.0155	0.5006	5.67	0.02
Isopentyl acetate *	7.8289	3.309	9.34	0.0032
2-Methyl-1-butanol	0.402	0.2769	1.07	0.3052
3-Methyl-1-butanol	0.5129	0.3514	0.45	0.5066
Ethyl hexanoate	0.2035	0.1724	0.24	0.6227
Hexyl acetate	0.0271	0.0262	0.03	0.8537
3-Hydroxy-2-butanone **	0.0279	0.0486	5.85	0.0183
cis-3-Hexenylacetate	0.0318	0.1708	3.51	0.0653
Hexanol	0.01	0.011	0.67	0.4175
Ethyl octanoate	0.2668	0.1358	0.89	0.3497
Benzaldehyde	0.1167	0.065	0	0.9476
Isovaleric acid *	1.1198	0.4681	11.69	0.0011
Diethyl succinate	0.1328	0.0899	0.52	0.4713
α-Terpineol	0.0025	0.0015	0.31	0.3341
Benzyl acetate	0.1321	0.0426	2.65	0.1081
Ethylphenyl acetate	1.4938	1.0218	3.77	0.0563
Hexanoic acid **	0.3401	0.1818	6.67	0.012
Phenylethanol *	1.5502	0.5302	8.91	0.0039

* Significant differences for $p < 0.01$. ** Significant differences for $p < 0.05$.

The orthogonal representation of the samples was used to observe the distribution of the data with regard to PC1 and PC2 (Figure 3). The distribution of the samples on this graphical representation did not show any clear differences between the vinegars from either origin, although Huelva vinegar samples did appear on the lower area of PC2.

The volatile compounds with a greater influence on PC1 were as follows: *n*-butyl acetate, 2-acetyl-5-methylfurfural, 4-ethylguaiacol, 4-ethylphenol, octanoic acid, decanoic acid, isopentyl acetate, 2-methyl-1-butanol, 3-methyl-1-butanol, ethyl octanoate, ethylphenyl acetate, and phenylethanol.

With regard to PC2, the compounds that demonstrated a greater difference between the samples were as follows: 4-ethylguaiacol, 4-ethylphenol, ethyl-2-methylbutyrate, hexanal, isovaleric acid, and hexanoic acid.

A backward stepwise linear discriminant analysis was carried out later on. In this case, the percentage of correctly classified samples was 100% for Huelva vinegars and 90.24% for Jerez vinegars. After calculating the discriminant function, only three of the samples were misclassified, with only one of the samples being made from Pedro Ximénez grapes. The percentage of samples correctly classified reached up to 94.12% altogether. The compounds that displayed coefficients with the greatest influence according to the discriminating function were 3-methyl-1-butanol, benzaldehyde, ethyl hexanoate, hexyl acetate, and 4-ethylguaiacol, among others. Other authors have successfully differentiated vinegar types, including Jerez and Huelva vinegars, according to their volatile content [17]. However, the volatile compounds with a greater influence on the discrimination in such studies were different from the ones obtained in the present study because these authors studied different volatile compounds.

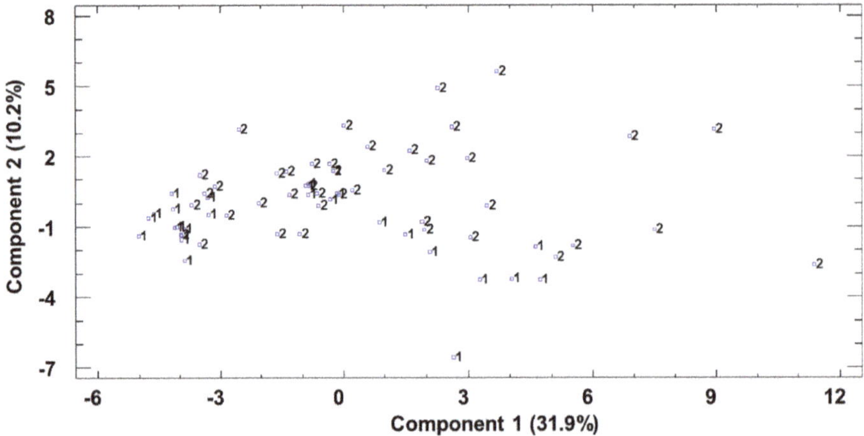

Figure 3. Principal component analysis of volatile compounds. Projection of the samples onto the plane formed by the first two principal components. 1—Huelva; 2—Jerez.

When the data on volatile compounds were subjected to cluster analysis using Ward's method and the Euclidean square distance metric as comparison criterion, the results failed to establish a clear difference between the two geographical indication, i.e., vinegars from both regions were arranged within the same groups, as can be seen in Figure 4.

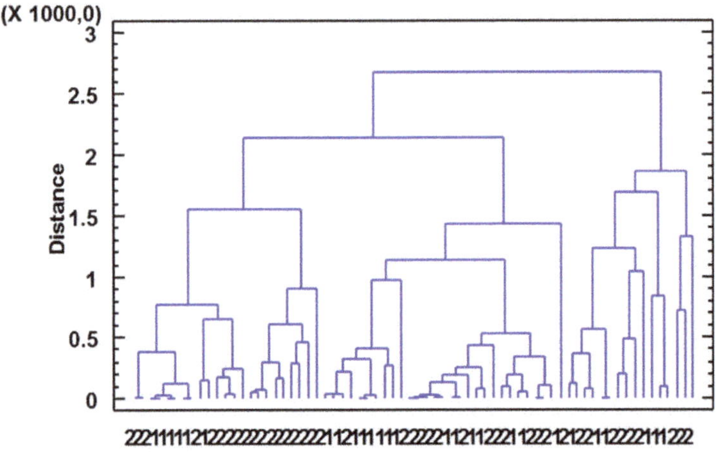

Figure 4. Cluster analysis of volatile compounds. 1—Huelva; 2—Jerez.

3.4. Joint Analysis of Polyphenolic, Furanic, and Volatile Constituents

Finally, in order to determine the most relevant variables for the differentiation of vinegars from the two PDOs, a principal component analysis was carried out in which the polyphenolic, furanic, and volatile constituents were taken into account.

In this case, 11 PCs (eigenvalue >1) were obtained to justify 88.8% of the samples' inherent variability. The first two PCs accounted for 53.5% of the variability. When the different samples were represented on the plane defined by these first two principal components (Figure 5), it could be observed that PC2 was the one that clearly differentiated Jerez vinegar (2) from Huelva ones (1), with negative values of this PC for Jerez vinegars.

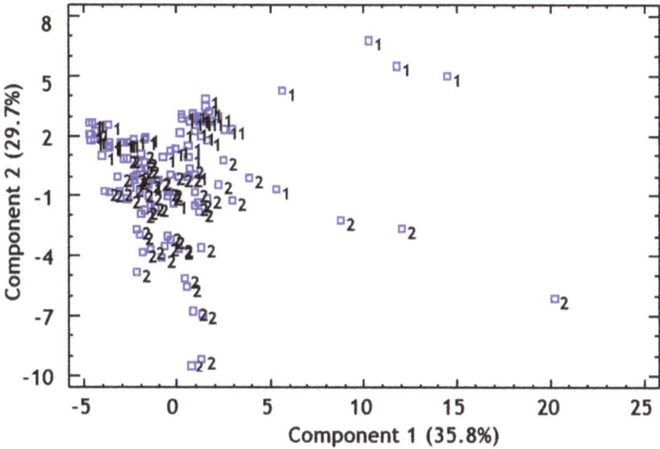

Figure 5. Principal component analysis of volatile, polyphenolic, and furanic compounds. Projection of the samples onto the plane formed by the first two principal components. 1—Huelva; 2—Jerez.

In this case, the polyphenolic and furanic compounds showed a clearly greater contribution to this principal component, with the hydroxycinnamic derivates (caftaric acid, *p*-coutaric acid, caffeic acid, *p*-coumaric acid, ethyl caffeate, and ethyl coumarate) as the variables with the greatest weight; all of them had a negative sign. With regard to the volatile compounds that contributed to this PC, 4-ethylguaiacol, propyl acetate, ethyl isobutyrate, ethyl isovalerate, 1-hexanol, and benzyl alcohol are worth mentioning.

4. Conclusions

The statistical study of polyphenols and volatile compounds in Jerez and Huelva PDO vinegars demonstrated clear differences between these two geographical indications. The differences were clearly based on their polyphenolic compound content. As vinegar manufacturing and ageing processes are similar in both regions, we should presume the importance of other factors, such as the grape varieties used as well as other geographical factors.

Author Contributions: Conceptualization, E.D.-G., R.C., and M.S.; methodology, E.D.-G. and M.S.; software, R.C.; validation, E.D.-G. and R.C.; formal analysis, E.D.-G. and M.S.; investigation, M.A.F.-R.; resources, C.G.B.; data curation, R.C.; writing—original draft preparation, E.D.-G.; writing—review and editing, E.D.-G., M.S., and R.C.; visualization, M.S. and M.A.F.-R.; supervision, R.C.; project administration, R.C.; funding acquisition, C.G.B.

Funding: This research received no external funding.

Conflicts of Interest: The authors declare no conflict of interest. The funders had no role in the design of the study; in the collection, analyses, or interpretation of data; in the writing of the manuscript; or in the decision to publish the results.

References

1. Solieri, L.; Giudici, P. Vinegars of the World. In *Vinegars of the World*; Springer Milan: Milano, Italy, 2009.
2. Chinnici, F.; Guerrero, E.D.; Sonni, F.; Natali, N.; Marín, R.N.; Riponi, C. Gas chromatography-mass spectrometry (GC-MS) characterization of volatile compounds in quality vinegars with protected European geographical indication. *J. Agric. Food Chem.* **2009**, *57*, 4784–4792. [CrossRef]
3. Junta de Andalucía. Inscripción de la Denominación de Origen Protegida "Vinagre de Jerez". Boletín Oficial de la Junta de Andalucía. 2008, pp. 29–35. Available online: https://www.juntadeandalucia.es/boja/2008/184/d15.pdf (accessed on 16 July 2019).
4. Junta de Andalucía. Inscripción de la Denominación de Origen Protegida "Vinagre del Condado de Huelva". In *Boletín Oficial de la Junta de Andalucía*; Junta de Andalucía: Sevillla, Spain, 2008; pp. 35–40.
5. Guerrero, E.D.; Castro Mejías, R.; Marín, R.N.; Barroso, C.G. Optimization of stir bar sorptive extraction applied to the determination of pesticides in vinegars. *J. Chromatogr. A.* **2007**, *1165*, 144–150. [CrossRef] [PubMed]
6. Callejón, R.M.; González, A.G.; Troncoso, A.M.; Morales, M.L. Optimization and validation of headspace sorptive extraction for the analysis of volatile compounds in wine vinegars. *J. Chromatogr. A.* **2008**, *1204*, 93–103. [CrossRef] [PubMed]
7. Guerrero, E.D.; Chinnici, F.; Natali, N.; Marín, R.N.; Riponi, C. Solid-phase extraction method for determination of volatile compounds in traditional balsamic vinegar. *J. Sep. Sci.* **2008**, *31*, 3030–3036. [CrossRef] [PubMed]
8. Dávalos, A.; Bartolomé, B.; Gómez-Cordovés, C. Antioxidant properties of commercial grape juices and vinegars. *Food Chem.* **2005**, *93*, 325–330. [CrossRef]
9. Garcia-Parrilla, C.; Heredia, F.J.; Troncoso, A.M. Phenols HPLC Analysis by Direct Injection of Sherry Wine Vinegar. *J. Liq. Chromatogr. Relat. Technol.* **1996**, *19*, 247–258. [CrossRef]
10. Cejudo Bastante, M.J.; Durán Guerrero, E.; Castro Mejías, R.; Natera Marín, R.; Rodríguez Dodero, M.C.; Barroso, C.G. Study of the polyphenolic composition and antioxidant activity of new sherry vinegar-derived products by maceration with fruits. *J. Agric. Food Chem.* **2010**, *58*, 11814–11820. [CrossRef] [PubMed]
11. Marrufo-Curtido, A.; Cejudo-Bastante, M.J.; Rodríguez-Dodero, M.C.; Natera-Marín, R.; Castro-Mejías, R.; García-Barroso, C.; Durán-Guerrero, E. Novel vinegar-derived product enriched with dietary fiber: Effect on polyphenolic profile, volatile composition and sensory analysis. *J. Food Sci. Technol.* **2015**, *52*, 7608–7624. [CrossRef] [PubMed]
12. Casale, M.; Armanino, C.; Casolino, C.; Oliveros, C.C.; Forina, M. A Chemometrical Approach for Vinegar Classification by Headspace Mass Spectrometry of Volatile Compounds. *Food Sci. Technol. Res.* **2006**, *12*, 223–230. [CrossRef]
13. Pizarro, C.; Esteban-Díez, I.; Sáenz-González, C.; González-Sáiz, J.M. Vinegar classification based on feature extraction and selection from headspace solid-phase microextraction/gas chromatography volatile analyses: A feasibility study. *Anal. Chim. Acta* **2008**, *608*, 38–47. [CrossRef] [PubMed]
14. Natera, R.; Castro, R.; De Valme García-Moreno, M.; Hernández, M.J.; García-Barroso, C. Chemometric studies of vinegars from different raw materials and processes of production. *J. Agric. Food Chem.* **2003**, *51*, 3345–3351. [CrossRef] [PubMed]
15. Cejudo-Bastante, M.J.; Durán-Guerrero, E.; Natera-Marín, R.; Castro-Mejías, R.; García-Barroso, C. Characterisation of commercial aromatised vinegars: Phenolic compounds, volatile composition and antioxidant activity. *J. Sci. Food Agric.* **2013**, *93*, 1284–1302. [CrossRef] [PubMed]
16. Marrufo-Curtido, A.; Cejudo-Bastante, M.J.; Durán-Guerrero, E.; Castro-Mejías, R.; Natera-Marín, R.; Chinnici, F.; García-Barroso, C. Characterization and differentiation of high quality vinegars by stir sorptive extraction coupled to gas chromatography-mass spectrometry (SBSE-GC-MS). *Lwt-Food Sci. Technol.* **2012**, *47*, 332–341. [CrossRef]
17. Ríos-Reina, R.; Segura-Borrego, M.P.; García-González, D.L.; Morales, M.L.; Callejón, R.M. A comparative study of the volatile profile of wine vinegars with protected designation of origin by headspace stir bar sorptive extraction. *Food Res. Int.* **2019**, *123*, 298–310. [CrossRef] [PubMed]
18. García-Parrilla, M.C.; González, G.A.; Heredia, F.J.; Troncoso, A.M. Differentiation of Wine Vinegars Based on Phenolic Composition. *J. Agric. Food Chem.* **1997**, *45*, 3487–3492. [CrossRef]

19. Guerrero, E.D.; Mejías, R.C.; Marín, R.N.; Lovillo, M.P.; Barroso, C.G. A new FT-IR method combined with multivariate analysis for the classification of vinegars from different raw materials and production processes. *J. Sci. Food Agric.* **2010**, *90*, 712–718. [CrossRef] [PubMed]

20. Ríos-Reina, R.; Callejón, R.M.; Oliver-Pozo, C.; Amigo, J.M.; García-González, D.L. ATR-FTIR as a potential tool for controlling high quality vinegar categories. *Food Control.* **2017**, *78*, 230–237. [CrossRef]

21. Ríos-Reina, R.; García-González, D.L.; Callejón, R.M.; Amigo, J.M. NIR spectroscopy and chemometrics for the typification of Spanish wine vinegars with a protected designation of origin. *Food Control.* **2018**, *89*, 108–116. [CrossRef]

22. Ríos-Reina, R.; Elcoroaristizabal, S.; Ocaña-González, J.A.; García-González, D.L.; Amigo, J.M.; Callejón, R.M. Characterization and authentication of Spanish PDO wine vinegars using multidimensional fluorescence and chemometrics. *Food Chem.* **2017**, *230*, 108–116. [CrossRef]

23. Úbeda, C.; del Barrio-Galán, R.; Peña-Neira, Á.; Medel-Maraboulí, M.; Durán-Guerrero, E. Location effects on the aromatic composition of monovarietal cv. Carignan wines. *Am. J. Enol. Vitic.* **2017**, *68*, 390–399. [CrossRef]

24. C.I.E. (Commission Internationale de L'Eclairage). Publication C.I.E. No.15.2. In *Colorimetry*; C.I.E.: Vienna, Austria, 1986.

25. Schwarz, M.; Rodríguez, M.C.; Guillén, D.A.; Barroso, C.G. Development and validation of UPLC for the determination of phenolic compounds and furanic derivatives in Brandy de Jerez. *J. Sep. Sci.* **2009**, *32*, 1782–1790. [CrossRef]

26. Guerrero, E.D.; Marín, R.N.; Mejías, R.C.; Barroso, C.G. Optimisation of stir bar sorptive extraction applied to the determination of volatile compounds in vinegars. *J. Chromatogr. A* **2006**, *1104*, 47–53. [CrossRef]

27. Kaiser, H.F. The Application of Electronic Computers to Factor Analysis. *Educ. Psychol. Meas.* **1960**, *20*, 141–151. [CrossRef]

Article

Physicochemical, Spectroscopic and Chromatographic Analyses in Combination with Chemometrics for the Discrimination of Four Sweet Cherry Cultivars Grown in Northern Greece

Spyridon Papapetros [1], Artemis Louppis [2], Ioanna Kosma [1], Stavros Kontakos [3], Anastasia Badeka [1], Chara Papastephanou [2] and Michael G. Kontominas [1,*]

[1] Laboratory of Food Chemistry, Department of Chemistry, University of Ioannina, 45110 Ioannina, Greece; spirosp32@gmail.com (S.P.); ikosma@cc.uoi.gr (I.K.); abadeka@uoi.gr (A.B.)

[2] cp Foodlab Ltd., Polifonti, 25, 2047 Strovolos, Nicosia, Cyprus; artemislouppis@gmail.com (A.L.); foodlab@cytanet.com.cy (C.P.)

[3] Department of Social Administration and Political Science, Democritus University of Thrace, 69100 Komotini, Greece; skon2915dt@gmail.com

* Correspondence: mkontomi@cc.uoi.gr

Received: 28 August 2019; Accepted: 22 September 2019; Published: 26 September 2019

Abstract: A total of 56 sweet cherry samples belonging to four cultivars (Ferrovia, Canada Giant, Lapins, and Germersdorfer) grown in northern Greece were characterized and differentiated according to botanical origin. For the above purpose, the following parameters were determined: conventional quality parameters (titratable acidity (TA), pH, total soluble solids (TSS), total phenolic content (TPC), mechanical properties and sensory evaluation, sugars by High Performance Liquid Chromatography (HPLC), volatile compounds by GC/MS, and minerals by ICP-OES. Statistical treatment of the data was carried out using Multivariate Analysis of Variance (MANOVA) and Linear Discriminant Analysis (LDA). The results showed that the combination of volatile compounds and conventional quality parameters provided a correct classification rate of 84.1%, the combination of minerals and conventional quality parameters 86.4%, and the combination of minerals, conventional quality parameters and sugars provided the highest correct classification rate of 88.6%. When the above four cherry cultivars were combined with previously studied Kordia, Regina, Skeena and Mpakirtzeika cultivars, collected from the same regions during the same seasons, the respective values for the differentiation of all eight cultivars were: 85.5% for the combination of conventional quality parameters, volatiles and minerals; and 91.3% for the combination of conventional quality parameters, volatiles, minerals, and sugars.

Keywords: sweet cherries; discrimination; physicochemical quality parameters; chemometrics

---

1. Introduction

Cherries have been cultivated for thousands of years in Europe, as pits have been recovered from cave dwellings dating back to 4000–5000 B.C. Today, cherry cultivation is very popular in Greece [1]. Sweet cherries (*Prunus avium L.*) are widely accepted for their quality characteristics, such as skin color, texture, sugar and organic acid content, and volatile compound composition [2].

Sweet cherries are a natural source of useful ingredients such as phenolic compounds functioning as natural antioxidants, which reduce the risk of degenerative diseases caused by oxidative stress; they are also a source of minerals, sugars, and organic acids [3–5].

Sugars are one of the main ingredients of sweet cherries, which, along with organic acids, lead to the unique balance of fruit flavor. The sugar content can be as high as one-quarter of the total weight of the fruit [6]. Of the different sugars normally found in sweet cherries (glucose, sucrose, fructose, maltose, and sorbitol), glucose and fructose account for approximately 90% of the total sugars in the fruit [7].

Sensory quality attributes of cherries include sweetness (sugar content), sourness (organic acid content), fruit weight, skin color, fruit firmness, and particular aroma. Usually aroma, making up a very small portion of fruit weight (0.01%–0.001% fresh weight basis), comprises of a mixture of a large number of volatile and semi-volatile compounds collectively contributing to fruit odor [8,9]. Classes of volatile aroma compounds include aldehydes, alcohols, esters, acids, terpenes, etc. Particular volatile compounds are characteristic of fruit cultivar and can, thus, be used for the differentiation of cherry cultivars.

Based on the above, the main objective of the present study was to characterize four popular cherry cultivars (Ferrovia, Canada Giant, Lapins, and Germersdorfer) grown in Northern Greece and to differentiate them according to cultivar. The Ferrovia cultivar was developed in the area of Puglia, Italy. It is cultivated in many geographical regions in Greece. The fruit is large, heart-shaped with a red, shiny skin, and a firm texture. The fruit flesh is pink in color, juicy, with a strong adherence to the stone. The flavor is of moderate sweetness [10]. The Canada Giant cultivar originates from Summerland, Canada. It gives a large, red, shiny skin, heart-shaped fruit of firm texture. Greece, due to its mild climate, has the comparative advantage of early sweet cherry ripening by 10–15 days as compared to the rest of Europe, with the exception of Turkey. The Lapins cultivar was developed by K.O. Lapin at the Agricultural Research Station in Summerland, British Columbia in Canada. This cherry cultivar gives a large, red, shiny skin, heart-shaped fruit of firm texture. The fruit flesh is very juicy and dark red in color [11]. The Germersdorfer cultivar is of Hungarian origin, grown in all central European countries with different names. It is a firm, crisp, juicy, aromatic, sweet, slightly acidic cherry variety with a bright red fruit color [1].

As of 1992, the European Union has set specific rules defining the status of products labeled as protected destination of origin (PDO), protected geographical indication (PGI), and traditional specially guaranteed (TSG). Such an act spurs the production of unique foods on the basis of their territorial, compositional and sensory traits, as well as their preparation methods. Such foodstuff is of higher value, both in domestic and international markets [12,13].

Techniques used for the authentication of foodstuff include HPLC, ICP-OES, GC/MS, IRMS, Electronic Nose, NMR, DNA analysis, etc. [14–17].

The present study comprises of the second half of the work recently published [17] by our group focusing on different four popular sweet cherry cultivars: Ferrovia, Canada Giant, Lapins, and Germersdorfer, also grown in northern Greece. Therefore, a second objective, considered a challenge given the large number of cherry cultivars studied all together, was to attempt to differentiate all eight cultivars, those studied in our previous work and those studied in the present work.

2. Materials and Methods

2.1. Samples

Fifty-six sweet cherry samples were collected from northern Greece (Edessa and neighbouring Kozani region) as follows: 19 (10 + 9, for each of two harvesting periods) samples of the Ferrovia cultivar, 12 (6 + 6) samples of the Canada Giant cultivar, 13 (7 + 6) samples of the Lapins cultivar, and 12 (6 + 6) samples of the Germersdorfer cultivar. Samples were collected during the period of 20 May–30 June for two consecutive years (2015–2016) at the stage of full ripeness, placed in rectangular plastic cups (500 g per cup), and transferred to the laboratory in insulated styrofoam boxes within four hours after collection.

2.2. Determination of Conventional Quality Parameters

Conventional quality parameters total soluble solids (TSS), pH, titratable acidity (TA), and mechanical properties were determined according to the methods described by Papapetros et al. [17]. TSS was measured by refractometry and expressed as Brix. TA was measured volumetrically and expressed as g of Malic acid equivalents (MAE) per 100 g Fresh Weight (FW).

Mechanical properties (force/load and penetration) were determined by dynamometry and expressed as (N) and (mm), respectively. TPC was determined according to Vavoura et al. [11] by adding the methanolic cherry extract to the Folin-Ciocalteu reagent and upon completion of the reaction, measurement of the absorbance at $\lambda = 725$ nm. Aqueous solutions of Gallic acid were used for quantification of TPC. Results were expressed as mg of Gallic acid equivalents/ 100 g FW.

Sensory evaluation was carried out according to Vavoura et al. [11] using a 51-member panel (acceptability test). Panelists evaluated color, texture and flavor on a 5-point hedonic scale with 5 corresponding to the most liked sample and 1 corresponding to the least liked sample. A score of 3 was taken as the lower limit of acceptability.

2.3. Determination of Sugars Using HPLC-RID

The main sugars in cherries are glucose and fructose. Sugar analysis, including method analytical characteristics, are described by Papapetros et al. [17]. All measurements were carried out in triplicate.

2.4. Identification and Semi-Quantification of Volatile Compounds Using Solid Phase Micro-Extraction in Combination with Gas Chromatography/Mass Spectrometry (SPME-GC/MS)

The analysis of volatile compounds was carried out using the method by Vavoura et al. [11].

2.5. Determination of Minerals Using Inductively Coupled Plasma Optical Emission Spectrometry (ICP-OES)

Elemental analysis was carried out according to the method described by Papapetros et al. [17].

2.6. Statistical Analysis

All measurements were carried out three-fold ($n = 3$) with the exception of mechanical properties carried out five-fold ($n = 5$). Analytical data were treated using the SPSS 23.0 Statistics software (IBM, Armonk, NY, USA) as described by Papapetros et al. [17].

3. Results and Discussion

3.1. Determination of Conventional Quality Parameters

The results for the conventional quality parameters are presented in Table 1. The values for cherry pH are quite similar, ranging from 3.98 ± 0.20 to 4.14 ± 0.09 for the Germersdorfer and the Canada Giant cultivars, respectively. Similar pH values were reported by Vursavuş et al. [18] for the three cheery cultivars: Van, Noir De Guben, and 0-900 Ziraat grown in Turkey (3.82, 4.20, and 4.10, respectively). Vavoura et al. [11] analysed cherries from four cultivars (Canada Giant, Lapins, Ferrovia, and Skeena) and recorded pH values of 3.91 ± 0.00, 3.96 ± 0.00, 3.81 ± 0.00 and 3.91 ± 0.00, respectively. Papapetros et al., [17] reported similar pH values for cherry cultivars: Kordia, Regina, Mpakirtzeika, and Skeena; with Regina recording the highest pH value (4.69). Tural and Koca [19] studied Cornelia cherry fruits and recorded lower pH values than those of the present study, ranging from 3.11 to 3.53. Likewise, Chockchaisawasdee et al. [20] observed variations in pH values among the 56 cherry cultivars analysed, ranging from 3.5 for the Van cultivar to 4.8 for the Minnulara cultivar.

TSS values of the tested cultivars ranged from 12.23 ± 1.99 Brix for the Lapins cultivar to 14.06 ± 2.27 Brix for the Ferrovia cultivar. Vursavuş et al. [18] recorded TSS values of 14.20% for the Van, 14.00% for the Noir De Guben, and 14.40% for the 0-900 Ziraat cultivar. On the other hand, Vavoura et al. [11] recorded higher TSS values for the cultivars: Canada Giant, Lapins, and Ferrovia

(14.50 ± 0.05, 13.00 ± 0.03, and 16.00 ± 0.06 Brix, respectively). Papapetros et al. [17] reported similar TSS values to those of the present study for cherry cultivars: Kordia, Regina, Mpakirtzeika, and Skeena, ranging between 12.8 and 14.6 Brix. Serradilla et al. [7] observed significant differences in TSS values, ranging from 13.97 Brix for the Sweetheart to 23.20 Brix for Pico Colorado. Likewise, Hayaloglu and Demir [21] investigated 12 cherry cultivars and recorded TSS values ranging from 13.71 ± 0.47 Brix for the Durona di Cesena to 19.55 ± 0.29 Brix for the Bing cherry cultivar. Tural and Koca [19] studied the Cornelia cherry cultivar and reported TSS values of 76.80–154.00 g/kg. For the Lapins cultivar, Hayaloglu and Demir [21] recorded a TSS value of 17.52 Brix, much higher than the Lapins TSS values of the present study.

TA was expressed as MAE per 100 g FW. TA values ranged from 0.27 ± 0.04 g MAE/100 g FW for the Germersdorfer cultivar to 0.35 ± 0.04 g MA/100 g FW for the Canada Giant. Lower TA values have been recorded by Vavoura et al. [11] for the four cultivars, Canada Giant, Lapins, Ferrovia, and Skeena (0.28 ± 0.01, 0.20 ± 0.00, 0.27 ± 0.00, and 0.21 ± 0.01 g MA/100g FW, respectively). Papapetros et al. [17] reported similar TA values to those of the present study for cherry cultivars: Kordia, Regina, Mpakirtzeika, and Skeena, ranging between 0.19 and 0.39 g MAE/100 g. Esti et al. [22] recorded higher TA values for the Ferrovia cultivar (0.81 ± 0.05 g/100 g) than those of the present study. Vursavuş et al. [18] observed differences in TA values of TA, ranging from 4.75 ± 0.01 g/L for the Noir De Gubento to 7.08 ± 0.01 g/L for the Van cultivar. According to Hayaloglu and Demir [21], TA recorded values between (0.71 ± 0.01) for the Durona di Cesena and (1.01 ± 0.01 g MA/100 g FW) for the Sweetheart cultivar, while Chockchaisawasdee et al. [20] reported similar TA values for the Pico Negro cultivar (0.6 g MA/100 g FW) and for the Black Star and Puntalazzese cultivars (1.3 g MA/100 g FW).

TPC was expressed as gallic acid equivalents (GAE) per 100 g. TPC recorded differences among cultivars ranging from 76.91 ± 41.83 mg GA/100 g for the Ferrovia cultivar to 132.55 ± 53.19 mg GA/100 g for the Germersdorfer cultivar. The TPC values reported for sweet cherries by Hayaloglu and Demir [21] were similar to those of the present study, ranging between 58.31 ± 10.56 for the Van cultivar to 115.41 ± 7.98 mg GAE/100 g FW for the Belge cultivar. Likewise, Chockchaisawasdee et al. [20] determined TPC values between 50 mg GAE/100 g for the Pico Colarado cultivar to 687.4 mg GAE/100 g for Santina cultivar, while the Ferrovia cultivar recorded higher TPC (97.4 mg GAE/100 g) than those of the present study. Usenik et al. [23] analyzed 13 sweet cherry cultivars from Slovenia and reported a TPC value of (44.3 ± 3.42 mg GAE/100 g FW) for the Lapins cultivar, lower than that of the present study. Finally, Vavoura et al. [11] reported TPC values generally higher than those of the present study, for the four cultivars, Canada Giant, Lapins, Ferrovia, and Skeena, ranging between 95.14 (Canada Giant) and 170.35 (Skeena) mg GAE/100 g FW.

In the penetration test, the Force (Load) showed a wide range of values i.e., 8.00 ± 1.30 N for the Canada Giant cultivar to 25.00 ± 12.10 N for the Germersdorfer cultivar. Esti et al. [22] ran the penetration test for the Ferrovia cultivar and recorded a value of 19.00 ± 0.10 N for Force (Load), similar to that recorded in the present study—17.90 ± 9.00 N. Papapetros et al., [17] reported similar Force (Load) values to those of the present study for cherry cultivars: Kordia, Regina, Mpakirtzeika, and Skeena, ranging between 11.3 and 20.6 N.

As far as sensory evaluation is concerned, all four cultivars showed no differences ($p > 0.05$) for fruit external color. Flesh color differed among cultivars with Lapins showing the darkest and Germersdorfer showing the lightest flesh color. Texture scores showed significant differences ($p < 0.05$) among cultivars tested, with Canada Giant and Germersdorfer showing the most acceptable texture. Likewise, taste differed ($p < 0.05$) for the samples of different cultivars and ranged between 3.89 ± 0.05 for the Ferrovia cultivar to 4.50 ± 0.00 for the Canada Giant cultivar, with the latter considered as having the most acceptable taste. This cultivar also showed the highest TA and given the fact that all cultivars had similar sugar content, it was most probably the balance between sugar/acidity of the Canada Giant cultivar that resulted in the highest preference by panelists. Dever et al. [24] performed sensory evaluation of 16 sweet cherry cultivars originating from three harvest periods (early, midseason, and late harvest). Among them, the Lapins cultivar (midseason harvest) recorded values of 7.2, 6.0, 3.2, 5.8 for the juiciness, sweetness, sourness, and flavor, respectively (10-point scoring scale). Serradilla et al. [7] studied the sourness, sweetness, and fruit flavor for four cherry cultivars. The values for the sourness ranged from 2.56 to 5.13 for Ambrunés and Sweetheart cultivars, respectively (10-point scoring scale). Sweetness recorded larger differences, with the Pico Colorado cultivar showing the lowest and the Ambrunés cultivar the highest value (3.26 and 6.31, respectively). Finally, the fruit flavor ranged among 4.25 for the Pico Negro cultivar to 6.23 for the Ambrunés cultivar. Finally, Vavoura et al. [11] reported significant differences in external and flesh color, texture/firmness, and flavor for the four cultivars, Canada Giant, Lapins, Ferrovia, and Skeena. Differences in sensory parameters between Vavoura et al. and the present study may be related to regional soil and environmental characteristics (Naousa vs. Kozani and Edessa).

Table 1. Mean values and standard deviations (SD) of conventional quality parameters and sugars of the cherry samples tested.

Cultivar	Ferrovia	Canada Giant	Lapins	Germersdorfer	p *
	Mean ± SD	Mean ± SD	Mean ± SD	Mean ± SD	
pH	4.08 ± 0.71 [a]	4.14 ± 0.09 [a]	4.05 ± 0.09 [a]	3.98 ± 0.20 [a]	0.951
TSS (Brix)	14.06 ± 2.27 [a]	13.11 ± 4.45 [a]	12.23 ± 1.99 [a]	12.33 ± 2.33 [a]	0.219
TA (gr MAE/100gr FW)	0.30 ± 0.04 [ab]	0.35 ± 0.04 [c]	0.31 ± 0.03 [b]	0.27 ± 0.04 [a]	0.001
TPC (mg GAE/100gr FW)	76.91 ± 41.83 [ab]	82.99 ± 8.60 [a]	128.6 ± 24.2 [bc]	132.6 ± 53.2 [c]	0.001
Fruit diameter (cm)	2.83 ± 0.23 [a]	2.58 ± 0.49 [a]	2.82 ± 0.16 [a]	2.84 ± 0.00 [a]	0.107
External color	4.87 ± 0.19 [a]	4.60 ± 0.00 [a]	4.54 ± 0.52 [a]	4.60 ± 0.00 [a]	0.057
Flesh color	4.25 ± 0.26 [ab]	4.30 ± 0.00 [ab]	4.53 ± 0.51 [b]	4.00 ± 0.00 [a]	0.001
Taste	3.89 ± 0.05 [a]	4.50 ± 0.00 [b]	3.95 ± 0.05 [a]	4.00 ± 0.00 [a]	0.001
Texture	4.00 ± 0.00 [a]	4.50 ± 0.00 [b]	4.00 ± 0.10 [a]	4.60 ± 0.00 [c]	0.000
Force/Load (N)	17.90 ± 9.00 [a]	8.00 ± 1.30 [a]	11.00 ± 2.80 [a]	25.00 ± 12.10 [a]	0.055
Penetration (mm)	10.37 ± 0.83 [c]	8.24 ± 0.33 [a]	9.08 ± 0.65 [b]	8.73 ± 0.63 [ab]	0.000
Glucose (g/100 g)	12.69 ± 0.08 [a]	12.01 ± 2.39 [a]	7.37 ± 1.20 [a]	16.25 ± 3.48 [a]	0.173
Fructose (g/100 g)	4.23 ± 1.55 [ab]	3.95 ± 0.62 [ab]	2.83 ± 0.32 [a]	4.62 ± 0.97 [b]	0.030

TSS: Total Soluble Solids; TA: Titratable Acidity; MAE: Malic Acid Equivalents; FW: Fresh Weight; TPC: Total Phenolic Content; GAE: Gallic Acid Equivalents; [a, b, c] Mean with different letters in the same row are significantly different; * p values are the result of the application of MANOVA ($p < 0.05$ statistically significant).

3.2. Analysis of Sugars

The results for cherry fructose and glucose content are shown in Table 1. More specifically, the Germersdorfer cultivar was found to be the richest in sugars among the four cultivars tested (16.25 ± 3.48 g/100 g for glucose and 4.62 ± 0.97 g/100 g for fructose). On the other hand, the Lapins cultivar had the lowest concentration for both sugars (7.37 ± 1.20 g/100 g glucose and 2.83 ± 0.32 g/100 g fructose). In all samples tested, glucose recorded a substantially higher concentration than that of fructose.

Papapetros et al. [17] reported similar values for glucose and fructose as those of the present study for cherry cultivars: Kordia, Regina, Mpakirtzeika, and Skeena, ranging between 9.2 and 17.9 g/100 g for glucose and between 2.1 and 5.1 g/100 g for fructose. Vursavuş et al. [18] determined the sugars: glucose, fructose, sucrose, and sorbitol in four sweet cherries cultivars, and reported a total amount of sugars equal to 103.87, 108.41, 108.88, and 113.13 g/kg of FW for Larian, Van, Noir de Guben, and 0-900 Ziraat, respectively. Of the sugars determined, glucose was found to have the highest content in cherry samples, followed by fructose, sorbitol, and sucrose. Usenik et al. [23] analyzed 13 cultivars of sweet cherries from Slovenia, including the Lapins cultivar, and reported that its concentration of glucose and fructose was 93.7 ± 3.13 g/kg FW and 79.9 ± 3.40 g/kg FW, respectively. The other tested cultivars recorded glucose content between 61.8 ± 6.67 g/kg FW (for Sylvia cultivar) and 123 ± 4.02 g/kg FW (for the Early Van Compact), while the fructose content ranged from 51.5 ± 5.68 g/kg FW (for the Ferprime cultivar) to 101.5 ± 5.14 g/kg FW (for Lala Star cultivar). Esti et al. [22] determined the conventional quality and sensorial changes in cherries of various cultivars after cool storage and reported high values for sugars, with fructose ranging from 4.8 ± 0.4 to 5.1 ± 0.4 g/100 g and glucose ranging from 5.8 ± 0.4 to 6.4 ± 0.4 g/100 g. In general, the sugar content of sweet cherry cultivars in the present work was of the same order of magnitude to that reported in the literature.

3.3. Volatile Compounds

Table 2 presents the groups of volatile compounds identified, including: aldehydes, alcohols, ketones, hydrocarbons and terpenes with aldehydes; ketones and alcohols were the most abundant classes of volatile compounds recorded. The major aldehydes were: acetaldehyde followed by (E)-2-hexenal and hexanal, known as green leaf volatiles and major contributors of cherry fruit flavor [25,26].

The Germersdorfer cultivar had the lowest concentration of aldehydes (0.070 ± 0.019 mg/kg), while the Ferrovia cultivar had the highest concentration (0.214 ± 0.058 mg/kg). Alcohol concentrations ranged from 0.122 ± 0.055 mg/kg in the Germersdorfer cultivar to 0.210 ± 0.102 mg/kg in the Lapins cultivar. From the group of alcohols, ethanol was the compound with the highest concentration in all four cultivars up to 0.113 ± 0.032 mg/kg for the Germersdorfer cultivar. Regarding ketones, the Germersdorfer cultivar recorded the highest concentration, 0.346 ± 0.239 mg/kg, and was the only cultivar in which both acetone and 2-butanone were identified. The last two categories of volatile compounds, i.e., hydrocarbons and terpenes, exhibited very low concentrations in all four cultivars.

Vavoura et al. [11] reported that carbonyl compounds were the most abundant volatile compounds, ranging from 14.75 μg/kg in the Lapins cultivar to 34.62 μg/kg in the Ferrovia cultivar. Alcohols gave the second-strongest signals, ranging from 5.56 for the Ferrovia cultivar to 22.21 μg/kg for the Skeena cultivar. According to Papapetros et al. [17], volatiles decreased in the following order: Skeena > Regina > Mpakirtzeika > Kordia cherry cultivar, with aldehydes being the most abundant class of volatile compounds, followed by alcohols.

Finally, according to Serradilla et al. [7], (E)-2-hexen-1-ol was the main alcohol present in Picato type and Sweetheart sweet cherries in Spain. However, in the present study, this compound was identified only in the Germersdorfer cultivar (0.001 ± 0.001 mg/kg).

Table 2. Mean values and SD (mg/kg) of volatile compounds of cherry samples tested.

Cultivars			Ferrovia	Canada Giant	Lapins	Germersdorfer	p ***
Volatiles	RI_{exp} *	RI_{lit} **	Mean ± SD	Mean ± SD	Mean ± SD	Mean ± SD	
Aldehydes							
Acetaldehyde	<500	<500	0.136 ± 0.045	0.174 ± 0.073	0.126 ± 0.009	0.046 ± 0.034	0.023
Hexanal	798	767	0.044 ± 0.013	0.017 ± 0.005	0.028 ± 0.013	0.013 ± 0.005	0.116
(E)-2-Hexenal	852	859	0.031 ± 0.012	0.009 ± 0.003	0.026 ± 0.012	0.008 ± 0.003	0.317
Butanal, 3-methyl-	614	605	0.003 ± 0.001	n.d.	0.002 ± 0.001	0.003 ± 0.001	0.736
Total			0.214 ± 0.058	0.200 ± 0.083	0.182 ± 0.055	0.070 ± 0.019	
Alcohols							
Ethanol	<500	<500	0.159 ± 0.059	0.203 ± 0.085	0.206 ± 0.120	0.113 ± 0.032	0.176
3-Buten-1-ol, 3-methyl-	727	728	0.003 ± 0.001	0.003 ± 0.001	0.004 ± 0.002	0.004 ± 0.002	0.488
2-Buten-1-ol, 3-methyl	777	768	0.005 ± 0.002	n.d.	n.d.	0.003 ± 0.001	0.320
(E)-2-Hexen-1-ol	862	859	n.d.	n.d.	n.d.	0.002 ± 0.001	0.001
Total			0.167 ± 0.078	0.206 ± 0.101	0.210 ± 0.102	0.122 ± 0.055	
Ketones							
2-Butanone	600	586	n.d.	n.d.	n.d.	0.004 ± 0.001	0.000
Acetone	<500	<500	0.128 ± 0.053	0.090 ± 0.045	0.230 ± 0.117	0.342 ± 0.145	0.000
Total			0.128 ± 0.091	0.090 ± 0.064	0.230 ± 0.163	0.346 ± 0.239	
Hydrocarbons							
Hexane	600	600	0.016 ± 0.006	0.018 ± 0.005	0.021 ± 0.008	0.002 ± 0.001	0.310
Cyclohexane	657	660	0.006 ± 0.002	0.012 ± 0.003	0.003 ± 0.001	n.d.	0.310
Heptane	700	700	0.005 ± 0.003	0.007 ± 0.004	0.009 ± 0.002	0.004 ± 0.001	0.123
Pentane, 2-methyl-	559	554	0.001 ± 0.000	0.008 ± 0.002	0.002 ± 0.001	n.d.	0.341
1,3-Pentadiene	516	501	0.005 ± 0.002	n.d.	n.d.	0.002 ± 0.001	0.124
1,3-Butadiene, 2-methyl-	520	502	0.006 ± 0.003	0.008 ± 0.003	0.001 ± 0.000	0.007 ± 0.002	0.016
Total			0.039 ± 0.005	0.053 ± 0.006	0.036 ± 0.008	0.015 ± 0.003	

Table 2. Mean values and SD (mg/kg) of volatile compounds of cherry samples tested.

Cultivars			Ferrovia	Canada Giant	Lapins	Germersdorfer	p ***
Volatiles	RI_{exp} *	RI_{lit} **	Mean ± SD	Mean ± SD	Mean ± SD	Mean ± SD	
Terpenes							
dl-Limonene	1035	1038	0.010 ± 0.006	0.032 ± 0.019	0.001 ± 0.000	0.018 ± 0.004	0.001
α-Terpinene	1019	1024	0.001 ± 0.000	0.004 ± 0.002	0.007 ± 0.003	n.d.	0.374
p-Cymene	1026	1033	0.015 ± 0.005	0.023 ± 0.006	0.014 ± 0.005	0.003 ± 0.001	0.576
α-Pinene	932	940	0.001 ± 0.000	0.008 ± 0.004	0.003 ± 0.001	n.d.	0.475
α-Thujene	923	929	0.002 ± 0.001	0.005 ± 0.002	0.002 ± 0.001	n.d.	0.430
β-Myrcene	994	986	0.003 ± 0.001	n.d.	n.d.	n.d.	0.530
1,8-Cineole	1033	1044	n.d.	n.d.	n.d.	0.001 ± 0.000	0.017
γ-Terpinene	1062	1065	0.005 ± 0.002	0.006 ± 0.003	n.d.	n.d.	0.510
Total			0.037 ± 0.002	0.078 ± 0.006	0.027 ± 0.002	0.022 ± 0.001	
Miscellaneous							
Acetic acid	614	605	0.002 ± 0.001	0.001 ± 0.000	0.003 ± 0.001	0.001 ± 0.001	0.813
Ethyl ether	<500	<500	0.005 ± 0.002	n.d.	n.d.	0.016 ± 0.005	0.147
Ethene, 1,1-dichloro-	540	510	0.017 ± 0.006	0.025 ± 0.008	0.043 ± 0.015	0.016 ± 0.006	0.383
Chloroform	622	618	0.016 ± 0.006	0.016 ± 0.006	0.020 ± 0.011	0.013 ± 0.003	0.228
Total			0.040 ± 0.008	0.042 ± 0.012	0.066 ± 0.020	0.046 ± 0.007	

* RIexp: experimental retention index, ** RIlit: literature retention index (NIST MS search), *** p values are the result of the application of MANOVA ($p < 0.05$ statistically significant), n.d. = not detected.

3.4. Minerals

All four cultivars had similar mineral concentrations. Mineral data are shown in Table 3. The highest amount of minerals was found in the Canada Giant cultivar, followed by the Ferrovia, Lapins, and Germersdorfer cultivars (2662 ± 437, 2452 ± 385, 2278 ± 347, and 2232 ± 351 mg/kg, respectively). The mineral identified with the highest value in the Canada Giant cultivar was Potassium. Phosphorous was the second-most abundant mineral, recording its highest concentration in the same cultivar (282.5 ± 52.1 mg/kg). Calcium and Magnesium both reported high concentrations in the Lapins cultivar (138.3 ± 60.9 and 134.4 ± 38.4 mg/kg, respectively). Minerals such as Be, Cr, Li, Se, Sn, Ti, Tl, and V were also identified, but in a very low concentration, lower than 1 mg/kg.

Table 3. Mean values and SD of minerals of cherry samples tested.

Cultivar	Ferrovia	Canada Giant	Lapins	Germersodfer	p *
	Mean ± SD	Mean ± SD	Mean ± SD	Mean ± SD	
Al	0.81 ± 0.50 [b]	0.59 ± 0.19 [ab]	1.06 ± 0.56 [b]	0.43 ± 0.22 [a]	0.055
B	5.22 ± 1.97 [a]	0.03 ± 0.01 [a]	4.73 ± 2.06 [a]	5.32 ± 1.53 [a]	0.621
Ba	0.49 ± 0.17 [a]	0.03 ± 0.01 [a]	0.25 ± 0.14 [a]	0.05 ± 0.01 [a]	0.770
Be	0.02 ± 0.01 [ab]	n.d.	0.03 ± 0.01 [b]	n.d.	0.001
Ca	130.3 ± 54.5 [b]	68.63 ± 13.88 [a]	138.3 ± 60.9 [b]	88.50 ± 16.00 [a]	0.018
Co	0.01 ± 0.00 [a]	0.02 ± 0.01 [a]	0.02 ± 0.01 [a]	0.01 ± 0.00 [a]	0.975
Cr	0.11 ± 0.08 [a]	0.09 ± 0.07 [a]	0.11 ± 0.07 [a]	0.12 ± 0.04 [a]	0.858
Cu	1.10 ± 0.28 [b]	1.34 ± 0.25 [b]	1.19 ± 0.30 [b]	0.87 ± 0.10 [a]	0.028
Fe	3.05 ± 0.84 [a]	3.29 ± 0.86 [a]	3.21 ± 1.79 [a]	2.34 ± 0.55 [a]	0.462
K	1922 ± 228 [a]	2186 ± 308 [b]	1729 ± 308 [a]	1753 ± 222 [a]	0.007
Li	0.04 ± 0.02 [a]	0.05 ± 0.01 [a]	0.02 ± 0.01 [a]	0.04 ± 0.02 [a]	0.104
Mg	127.4 ± 17.9 [a]	113.4 ± 23.7 [a]	134.4 ± 38.4 [a]	135.6 ± 30.2 [a]	0.435
Mn	1.42 ± 1.01 [a]	0.94 ± 0.12 [a]	1.89 ± 0.92 [a]	0.59 ± 0.23 [a]	0.138
Mo	0.01 ± 0.00 [ab]	n.d.	0.03 ± 0.01 [b]	n.d.	0.031
Ni	0.06 ± 0.03 [a]	0.07 ± 0.01 [a]	0.09 ± 0.03 [a]	0.10 ± 0.05 [a]	0.588
P	255.5 ± 40.8 [a]	282.5 ± 52.1 [a]	260.0 ± 87.0 [a]	241.4 ± 56.0 [a]	0.698
Sb	0.18 ± 0.07 [a]	0.21 ± 0.05 [a]	0.16 ± 0.07 [a]	0.19 ± 0.04 [a]	0.444
Se	0.01 ± 0.00 [a]	n.d.	0.01 ± 0.00 [a]	n.d.	0.452
Si	1.99 ± 0.91 [a]	3.07 ± 4.36 [a]	1.50 ± 1.20 [a]	1.72 ± 0.43 [a]	0.359
Sn	0.34 ± 0.17 [a]	0.35 ± 0.09 [a]	0.25 ± 0.08 [a]	0.31 ± 0.04 [a]	0.918
Sr	0.24 ± 0.09 [b]	0.14 ± 0.06 [a]	0.33 ± 0.13 [b]	0.22 ± 0.08 [ab]	0.011
Ti	0.01 ± 0.00 [ab]	n.d.	0.04 ± 0.05 [b]	n.d.	0.008
Tl	0.47 ± 0.14 [a]	0.58 ± 0.17 [a]	0.47 ± 0.24 [a]	0.44 ± 0.11 [a]	0.511
V	0.03 ± 0.01 [a]	0.07 ± 0.03 [b]	0.03 ± 0.01 [a]	0.10 ± 0.02 [b]	0.000
Zn	1.28 ± 1.05 [a]	0.79 ± 0.23 [a]	1.12 ± 0.35 [a]	0.87 ± 0.35 [a]	0.673
Total	2452 ± 385	2662 ± 437	2278 ± 347	2232 ± 351	

[a, b] Means with different letters in the same row are significantly different; * p values are the result of the application of MANOVA ($p < 0.05$ statistically significant), n.d. = not detected.

There are only a few studies in the literature reporting mineral content in cherries. De Souza et al. [27] determined five minerals (P, K, Zn, Mg, Fe) in cherries from Brazil. Potassium was the main mineral (highest concentration equal to 90.92 mg/100 g FW), followed by P and Mg with similar concentrations (12.2 to 12.3 mg/100 g FW), Fe (1.16 mg/100 g FW), and Zn (0.69 mg/100 g FW). It should be noted that Ca was not detected in any of the samples analyzed. The range of concentrations for the main minerals in the above study is similar to those in the present study with the exception of Ca, which was not identified in the above study. Papapetros et al. [17] reported similar mineral content for cherry cultivars: Kordia, Regina, Mpakirtzeika, and Skeena, ranging between 2114 mg/kg for the Kordia cultivar and 2520 mg/kg for the Skeena cultivar. Finally, Matos-Reyes et al. [15] determined minerals in various cherry cultivars grown in Spain. The mineral showing the highest concentration was K ranging from 13,000 mg/kg in samples from Cáceres to 5500 mg/kg in Aragón cherries, followed by Ca and Mg present in concentrations higher than 500 mg/kg. Sodium varied from 10 mg/kg in Huesca sample to 70 mg/kg in Aragón and Castellón samples. The rest of the minerals were present in concentrations lower than 1 mg/kg.

3.5. Cultivar Differentiation of Four Cherry Cultivars (Ferrovia, Canada Giant, Lapins, and Germersdorfer) Based on Analytical Parameters

The 56 cherry samples were subjected to MANOVA in order to determine those parameters that are significant for the differentiation of cultivars. Dependent variables initially included the 28 volatile compounds, while cultivar was taken as the independent variable [28]. Pillai's Trace = 2.619 (F = 2.750, p-value = 0.001 < 0.05) and Wilks' Lambda = 0.001 (F = 2.763, p-value = 0.001 < 0.05) index values showed the existence of a significant multivariable effect of cultivar origin on the identity of cherry volatile compounds. Seven of the 28 volatile compounds were found to be significant (p < 0.05) for the differentiation of cherries according to cultivar and thus, were subjected to LDA. In LDA analysis, the cultivar was taken as the dependent variable, while the measured physicochemical parameters were taken as the independent variables [28]. The overall correct classification rate was 77.3% using the original and 65.9% using the cross-validation method, not a very satisfactory rate.

Similar statistical treatment was used for the conventional quality parameters and minerals, which gave a respective correct classification rate of 72.7% and 75%. Sugar statistical analysis showed that only fructose was significant for the differentiation of cultivars and thus, the formation of only one discriminant function showed that the statistical model developed was unable to provide results regarding the differentiation of cherry cultivar. In order to increase the correct classification rate, combinations of analytical sets of data were tested.

The combination of volatile compounds and conventional quality parameters were taken as the dependent variables, while cultivar was taken as the independent variable. Pillai's Trace = 2.920 (F = 4.924, p-value = 0.001 < 0.05) and Wilks' Lambda = 0.001 (F = 5.280, p-value = 0.001 < 0.05) index values showed that there is a significant multivariate effect of volatile compounds and conventional quality parameters on cherry cultivar. Seven of the volatile compounds and six conventional quality parameters were found to be significant (p < 0.05) for the differentiation of cultivar. These were then subjected to LDA. The results of statistical treatment are shown in Table 4. In Figure 1, it is shown that all cultivars are well differentiated, while the Lapins and the Canada Giant are quite close to each other. The overall correct classification rate achieved was 97.7% for the original, while for the cross-validation method, the respective rate was 84.1%, very satisfactory for both methods.

Table 4. Discriminant functions formed and MANOVA results for each function for parameter combinations tested.

Parameters Combinations	Discriminant Function	% of Variance	% Total Variance	Wilks' Lambda	X^2	df	p
Volatile Compounds-Conventional Quality Parameters (four cultivars)	1	51.1	51.1	0.012	152.171	39	0.001
	2	33.9	85.0	0.080	87.024	24	0.001
Minerals—Conventional Quality Parameters (four cultivars)	1	57.3	57.3	0.014	149.262	36	0.001
	2	25.2	82.5	0.098	81.361	22	0.001
Minerals—Conventional Quality Parameters—Sugars (four cultivars)	1	55.4	55.4	0.013	150.892	39	0.001
	2	27.1	82.5	0.088	83.858	24	0.001
Minerals—Conventional Quality Parameters—Volatile compounds (eight cultivars)	1	42.4	42.4	0.001	923.667	259	0.001
	2	38.7	81.1	0.001	712.119	216	0.001
	3	6.5	87.6	0.003	504.682	175	0.000
Minerals—Conventional Quality Parameters—Sugars—Volatile compounds (eight cultivars)	1	51.9	51.9	0.001	971.743	266	0.001
	2	26.8	78.7	0.001	742.568	222	0.001
	3	10.5	89.2	0.005	542.909	180	0.001

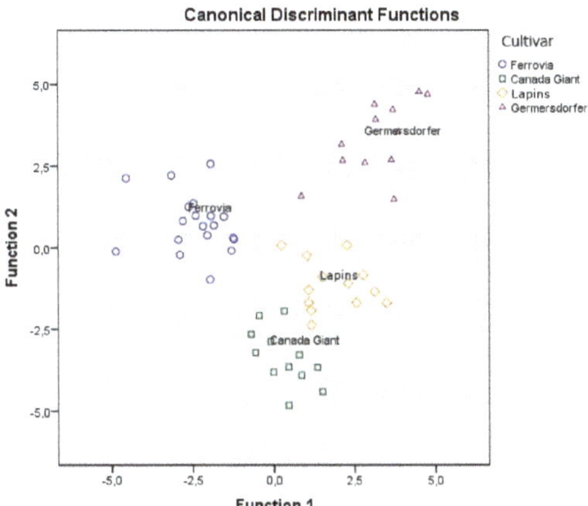

Figure 1. Differentiation of four cultivars based on the combination of volatile compounds and conventional quality parameters.

Likewise, the same statistical treatment was applied to the other combinations of sets of analytical data. Statistical analysis of minerals and conventional quality parameters showed that only eight minerals and six of the conventional quality parameters were found to be significant ($p < 0.05$) for cultivar differentiation (Table 4). The overall correct classification rate achieved was 97.7% for the original and 86.4% for the cross-validation method. In Figure 2, it is obvious that Ferrovia and Canada Giant are well differentiated from the other cultivars.

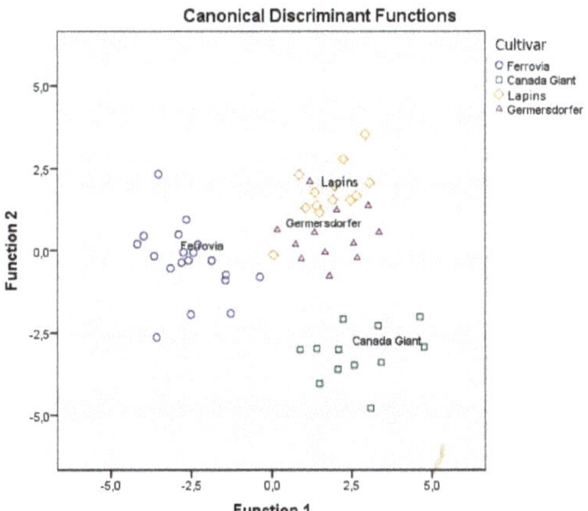

Figure 2. Differentiation of four cultivars based on the combination of minerals and conventional quality parameters.

Despite the fact that sugars per se could not provide information on the differentiation of cultivars, their combination with minerals and conventional quality parameters showed that eight minerals, six of the conventional quality parameters, and fructose were found to be significant ($p < 0.05$) for cultivar differentiation (Table 4). The overall correct classification rate achieved was 100% for the original and 88.6% for the cross-validation method, a very satisfactory value for both methods. In Figure 3, it is shown that Ferrovia and Canada Giant are very well differentiated, while Germersdorfer and Lapins are reasonably close to each other. Matos-Reyes et al. [15] determined the mineral content of Spanish cherries from different geographical areas (Aragón, Cáceres, Castellón, Huesca, and Alicante's

Mountain) using ICP-OES. Of the 42 elements determined, only 22 and 23 minerals for stones and fruit edible part were shown to be significant for the differentiation of cherry cultivars, respectively. The classification of cherry stones and edible parts was carried out using LDA. The results showed a correct classification rate of 100% for the edible part of cherries and 96.43% for the stone part. Unfortunately, no mention was made in this study on the cherry cultivars used. Finally, in a similar work, Papapetros et al. [17] reported a correct classification rate equal to 82.1% based on minerals, 89.5% based on conventional parameters, and 89.7% based on volatile compounds for the differentiation of Regina, Kordia, Mpakirtzeika, and Skeena cherry cultivars.

Other combinations tested that gave lower correct classification rates were volatiles and sugars (77.1%), volatiles and minerals (77.3%), conventional quality parameters and sugars (77.5%).

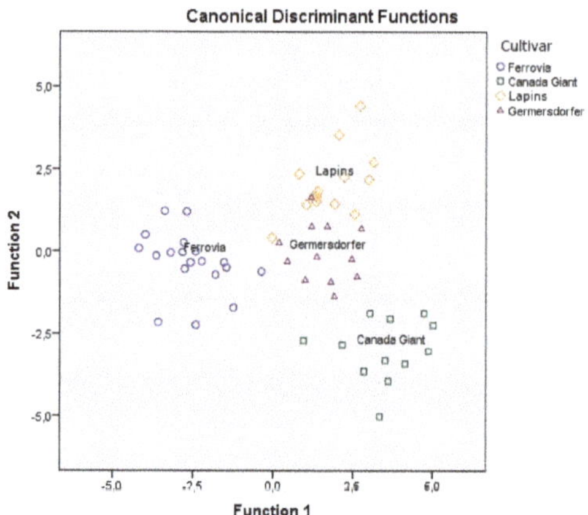

Figure 3. Differentiation of four cultivars based on the combination of minerals, conventional quality parameters, and sugars.

3.6. Cultivar Differentiation of All Eight Cherry Cultivars (Ferrovia, Canada Giant, Lapins, Germersdorfer, Kordia, Regina, Skeena, And Mpakirtzeika) Based on Analytical Parameters

Similarly, as described above, all 108 cherry samples were subjected to MANOVA in order to determine those volatile compounds that are significant for the differentiation of cultivars. Nineteen of the thirty volatile compounds were found to be significant ($p < 0.05$) for the differentiation of cherries according to cultivar and thus, were subjected to LDA. The overall correct classification rate was 89.9% using the original and 69.6% using the cross-validation method, not a very satisfactory rate.

Similar statistical treatment was used for the conventional quality parameters and minerals, which gave a respective correct classification rate of 70% and 61.4%. As already stated above, sugar statistical analysis showed that only fructose was significant for the differentiation of cultivars and thus, the formation of only one discriminant function showed that the statistical model developed was unable to provide results regarding the differentiation of cherry cultivar. In order to increase the correct classification rate, combinations of analytical sets of data were tested.

When volatile compounds and conventional quality parameters were combined, the Pillai's Trace and Wilks' Lambda index values showed a significant multivariate effect of volatile compounds and conventional quality parameters on cherry cultivar. Nineteen of the volatile compounds and five of the conventional quality parameters were found to be significant ($p < 0.05$) for the differentiation of cultivar. These were then subjected to LDA. Results showed an overall correct classification rate of 98.5% for the original and 85.3% for the cross-validation method, very satisfactory for both methods.

Likewise, the same statistical treatment was applied to the other combinations of sets of analytical data. Statistical analysis of i.e., minerals and conventional quality parameters, showed that only 13 minerals and 5 of the conventional quality parameters were found to be significant ($p < 0.05$) for cultivar differentiation. The overall correct classification rate achieved was 100% for the original and 85.1% for the cross-validation method. In terms of classification rate, the two most successful combinations were: (i) conventional quality parameters plus volatiles plus minerals resulting in an overall correct classification rate of 100% for the original, and 85.5% for the cross validation method, indeed a very satisfactory value for both methods (Table 4, Figure 4a,b); and (ii) the combination: conventional quality parameters, volatiles, minerals, and sugars (fructose), resulting in an overall correct classification rate of 100% for the original and 91.3% for the cross-validation method (Table 4, Figure 5a,b).

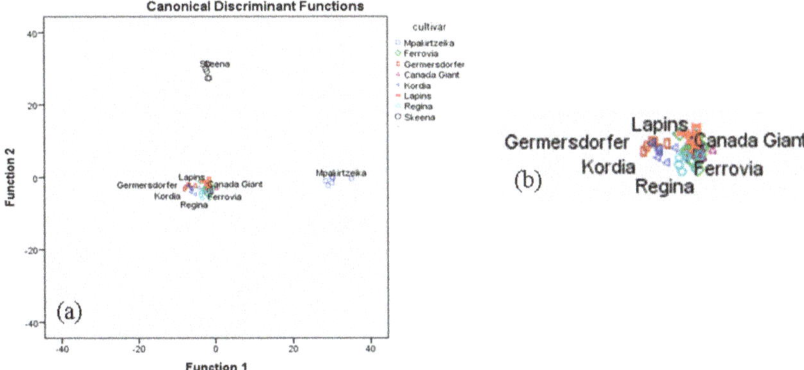

Figure 4. (**a**) Differentiation of eight cultivars based on the combination of conventional quality parameters, volatiles, and minerals; (**b**) blow up of Figure 4a.

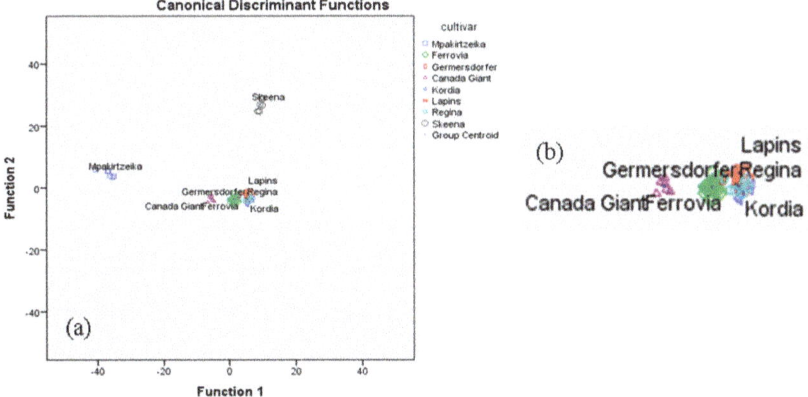

Figure 5. (**a**) Differentiation of eight cultivars based on the combination of conventional quality parameters, volatiles, minerals, and sugars; (**b**) blow up of Figure 5a.

In Figure 4a, it is shown that the Skeena and Mpakirtzeika cultivars are very well differentiated from the rest. In Figure 4b, it is shown that the Germersdorfer cultivar is well differentiated from the Regina, Lapins, Ferrovia, and Canada Giant cultivars, but not from the Kordia cultivar. Lapins, Ferrovia, and Canada Giant are considerably overlapping.

In Figure 5a, it is shown that the Skeena and Mpakirtzeika cultivars are again very well differentiated from the rest. In Figure 5b, it is shown that the Canada Giant cultivar is well differentiated from the Ferrovia, Lapins, Regina and Kordia cultivars, but not from the Germersdorfer cultivar. Lapins and Regina as well as Regina and Kordia are considerably overlapping.

4. Conclusions

Analysis of volatile compounds, minerals, and conventional quality parameters showed significant differences among cherry cultivars tested. Statistical treatment of the individual sets of data gave acceptable but not satisfactory correct classification rate (volatile compounds: 65.9%, conventional quality parameters: 72.7% and minerals: 75%). Furthermore, combinations of selected data sets increased correct classification rate i.e., volatiles and conventional quality parameters: 84.1%, minerals and conventional quality parameters: 86.4%. Finally, even though sugars per se could not provide information on cherry cultivar differentiation, when combined with minerals and conventional quality parameters, the classification rate was increased to 88.6%. This was the highest rate achieved in the present study, suggesting that the use of multi-element analysis may be a useful tool for cherry cultivar differentiation. In our previous work, Papapetros et al. [17], we achieved a very satisfactory differentiation (97.4%) of the botanical origin of four cherry cultivars (Kordia, Regina, Skeena, and Mpakirtzeika) grown in northern Greece using the same analytical methodology. In the present study, in a similar attempt, we were able, for the first time, to successfully differentiate (classification rate 88.6%) the botanical origin of four additional popular cherry cultivars (Ferrovia, Canada Giant, Lapins, and Germersdorfer) grown in the same greater area and during the same seasons in Greece. Finally, differentiation of all eight cherry cultivars was achieved with a classification rate of 85.5% for the combination of conventional quality parameters, volatiles, and minerals; and 91.3% for the combination of conventional quality parameters, volatiles, minerals, and sugars.

Author Contributions: Conceptualization, M.G.K.; Methodology, S.P.; I.K.; A.L.; Software, A.B.; C.P.; S.K.; Validation, S.P.; I.K.; A.L.; Formal analysis, S.P., I.K., A.L.; Investigation, S.P.; I.K., Data interpretation, S.P.; I.K.; A.L.; A.B.; S.K.; Writing—original draft preparation, S.P.; I.K.; A.B.; M.G.K., Writing—review and editing, M.G.K., Visualization M.G.K., Supervision, M.G.K.; A.B., Project administration, M.G.K.

Funding: This research received no external funding.

Acknowledgments: The authors would like to thank the cherry growers of Edessa and Kozani for providing samples.

Conflicts of Interest: The authors declare no conflict of interest.

References

1. Kazantzis, K.A. Description of the Main Cherry Cultivars in Greece. Available online: http://blog.farmacon.gr/katigories/texniki-arthrografia/pollaplasiastiko-yliko/item/426-perigrafi-kyrioteron-poikilion-kerasias-sti-xora-mas (accessed on 26 July 2016).

2. Diaz-Mula, H.M.; Castillo, S.; Martinez-Romero, D.; Valero, D.; Zapata, P.J.; Guillen, F.; Serrano, M. Sensory, nutritive and functional properties of sweet cherry as affected by cultivar and ripening stage. *Food Sci. Technol. Int.* **2009**, *15*, 535–543. [CrossRef]

3. Kim, D.O.; Heo, H.J.; Kim, Y.J.; Yang, H.S.; Lee, C.Y. Sweet and sour cherry phenolics and their protective effects on neuronal cells. *J. Agric. Food Chem.* **2005**, *53*, 9921–9927. [CrossRef] [PubMed]

4. Serrano, M.; Guillén, F.; Martínez-Romero, D.; Castillo, S.; Valero, D. Chemical constituents and antioxidant activity of sweet cherry at different ripening stages. *J. Agric. Food Chem.* **2005**, *53*, 2741–2745. [CrossRef] [PubMed]

5. Mulabagal, V.; Lang, G.A.; DeWitt, D.L.; Dalavoy, S.S.; Nair, M.G. Anthocyanin content, lipid peroxidation and cyclooxygenase enzyme inhibitory activities of sweet and sour cherries. *J. Agric. Food Chem.* **2009**, *57*, 1239–1246. [CrossRef] [PubMed]

6. Girard, B.; Kopp, T.G. Physicochemical characteristics of selected sweet cherry cultivars. *J. Agric. Food Chem.* **1998**, *46*, 471–476. [CrossRef] [PubMed]

7. Serradilla, M.J.; Martín, A.; Ruiz-Moyano, S.; Hernández, A.; López-Corrales, M.; Córdoba, M.D.G. Physicochemical and sensorial characterisation of four sweet cherry cultivars grown in Jerte Valley (Spain). *Food Chem.* **2012**, *133*, 1551–1559. [CrossRef]

8. Zhang, X.; Jiang, Y.M.; Peng, F.T.; He, N.B.; Li, Y.J.; Zhao, D.C. Changes of aroma components in Hongdeng sweet cherry during fruit development. *Agric. Sci. China* **2007**, *6*, 1376–1382. [CrossRef]

9. Xue, J.; Tu, Z.; Chang, W.; Jia, S.; Wang, Y. Analysis of the characteristic aroma components of wild fruit China dwarf cherry GC–MS. *J. Chin. Inst. Food Sci. Technol.* **2008**, *1*, 125–129.

10. Kazantzis, K.A. *Monograph on Sweet Cherry Cultivars*; Greek Agriculture Organization DEMETRA: Athens, Greece, 2013.

11. Vavoura, M.V.; Badeka, A.V.; Kontakos, S.; Kontominas, M.G. Characterization of Four Popular Sweet Cherry Cultivars Grown in Greece by Volatile Compound and Physicochemical Data Analysis and Sensory Evaluation. *Molecules* **2015**, *20*, 1922–1940. [CrossRef]

12. European Union. 2081/92 of 14 July 1992 on the protection of geographical indications and designations of origin for agricultural products and food stuffs. *Off. J. Eur. Union* **1992**, *208*, 1–8.

13. European Union. 2082/92 of 14 July 1992 on certificates of specific character for agricultural products and food stuffs. *Off. J. Eur. Union* **1992**, *208*, 9–14.

14. Pasqualone, A. Cultivar identification and varietal traceability in processed foods: A molecular approach. In *Cultivars: Chemical Properties, Antioxidant Activities and Health Benefits*; Carbone, K., Ed.; Nova Science Publishers: New York, NY, USA, 2013; pp. 83–106.

15. Matos-Reyes, M.N.; Simonot, J.; López-Salazar, O.; Cervera, M.L.; De la Guardia, M. Authentication of Alicante's Mountain cherries protected designation of origin by their mineral profile. *Food Chem.* **2013**, *141*, 2191–2197. [CrossRef] [PubMed]

16. Longobardi, F.; Casiello, G.; Ventrella, A.; Mazzilli, V.; Nardelli, A.; Sacco, D.; Catucci, L.; Agostiano, A. Electronic nose and isotope ratio mass spectrometry in combination with chemometrics for the characterization of the geographical origin of Italian sweet cherries. *Food Chem.* **2015**, *170*, 90–96. [CrossRef] [PubMed]

17. Papapetros, S.; Louppis, A.; Kosma, I.; Kontakos, S.; Badeka, A.; Kontominas, M.G. Characterization and differentiation of botanical and geographical origin of selected popular sweet cherry cultivars grown in Greece. *J. Food Compos. Anal.* **2018**, *72*, 48–56. [CrossRef]

18. Vursavuş, K.; Kelebek, H.; Selli, S. A study on some chemical and physico-mechanic properties of three sweet cherry varieties (*Prunus avium* L.) in Turkey. *J. Food Eng.* **2006**, *74*, 568–575. [CrossRef]

19. Tural, S.; Koca, I. Physicochemical and antioxidant properties of Cornelian cherry fruits (*Cornus mas* L.) grown in Turkey. *Sci. Hortic.* **2008**, *116*, 362–366. [CrossRef]

20. Chockchaisawasdee, S.; Golding, J.B.; Vuong, Q.V.; Papoutsis, K.; Stathopoulos, C.E. Sweet cherry: Composition, postharvest preservation, processing and trends for its future use. *Trends Food Sci. Technol.* **2016**, *55*, 72–83. [CrossRef]

21. Hayaloglu, A.A.; Demir, N. Physicochemical Characteristics, Antioxidant Activity, Organic Acid and Sugar Contents of 12 Sweet Cherry (*Prunus avium* L.) Cultivars Grown in Turkey. *J. Food Sci.* **2015**, *80*, 564–570. [CrossRef]

22. Esti, M.; Cinquanta, L.; Sinesio, F.; Moneta, E.; Di Matteo, M. Physicochemical and sensory fruit characteristics of two sweet cherry cultivars after cool storage. *Food Chem.* **2002**, *76*, 399–405. [CrossRef]

23. Usenik, V.; Fabčič, J.; Štampar, F. Sugars, organic acids, phenolic composition and antioxidant activity of sweet cherry (*Prunus avium* L.). *Food Chem.* **2008**, *107*, 185–192. [CrossRef]

24. Dever, M.C.; MacDonald, R.A.; Cliff, M.A.; Lane, W.D. Sensory Evaluation of Sweet Cherry Cultivars. *HortScience* **1996**, *31*, 150–153. [CrossRef]

25. Schmid, W.; Grosch, W. Identification of volatile aromas with high aromas in sour cherries (Prunus cerasus L.). *Z. Lebensm. Unters. Forsch.* **1986**, *182*, 407–412. [CrossRef]

26. Matsui, K. Green leaf volatiles: Hydroperoxidelyase pathway of oxylipin metabolism. *Curr. Opin. Plant Biol.* **2006**, *9*, 274–280. [CrossRef] [PubMed]

27. De Souza, V.R.; Pereira, P.A.P.; Da Silva, T.L.T.; De Oliveira Lima, L.C.; Pio, R.; Queiroz, F. Determination of the bioactive compounds, antioxidant activity and chemical composition of Brazilian blackberry, red raspberry, strawberry, blueberry and sweet cherry fruits. *Food Chem.* **2014**, *156*, 362–368. [CrossRef] [PubMed]

28. Field, A. *Discovering Statistics Using SPSS*; Sage Publications Ltd.: London, UK, 2009.

Article

A Robust DNA Isolation Protocol from Filtered Commercial Olive Oil for PCR-Based Fingerprinting

Luciana Piarulli [1,†]**, Michele Antonio Savoia** [1,†]**, Francesca Taranto** [1,2,*]**, Nunzio D'Agostino** [3]**, Ruggiero Sardaro** [4]**, Stefania Girone** [1]**, Susanna Gadaleta** [5]**, Vincenzo Fucili** [4]**, Claudio De Giovanni** [5]**, Cinzia Montemurro** [1,5,*]**, Antonella Pasqualone** [5] **and Valentina Fanelli** [5]

[1] SINAGRI S.r.l.—Spin Off of the University of Bari Aldo Moro, 70126 Bari, Italy; luciana.piarulli@libero.it (L.P.); micsav87@gmail.com (M.A.S.); stefaniagirone@hotmail.com (S.G.)
[2] CREA Research Centre for Cereal and industrial Crops (CREA-CI) S.S. 673, km 25.200, 71122 Foggia, Italy
[3] Department of Agricultural Sciences, University of Naples Federico II, via Università 100, 80055 Portici, Napoli, Italy; nunzio.dagostino@unina.it
[4] Department of Agricultural and Environmental Sciences, University of Bari Aldo Moro, 70126 Bari, Italy; ruggierosardaro@gmail.com (R.S.); vincenzo.fucilli@uniba.it (V.F.)
[5] Department of Soil, Plant and Food Sciences, University of Bari Aldo Moro, 70126 Bari, Italy; sanna14@hotmail.it (S.G.); claudio.degiovanni@uniba.it (C.D.G.); antonella.pasqualone@uniba.it (A.P.); valentina.fanelli@uniba.it (V.F.)
* Correspondence: francesca.taranto@uniba.it (F.T.); cinzia.montemurro@uniba.it (C.M.); Tel.: +39-0881-714-911 (ext. 435) (F.T.); +39-0805-443-003 (C.M.)
† Luciana Piarulli and Michele Antonio Savoia contributed equally to the article.

Received: 29 August 2019; Accepted: 6 October 2019; Published: 9 October 2019

Abstract: Extra virgin olive oil (EVOO) has elevated commercial value due to its health appeal, desirable characteristics and quantitatively limited production, and thus it has become an object of intentional adulteration. As EVOOs on the market might consist of a blend of olive varieties or sometimes even of a mixture of oils from different botanical species, an array of DNA-fingerprinting methods have been developed to check the varietal composition of the blend. Starting from a comparison between publicly available DNA extraction protocols, we set up a timely, low-cost, reproducible and effective DNA isolation protocol, which allows an adequate amount of DNA to be recovered even from commercial filtered EVOOs. Then, in order to verify the effectiveness of the DNA extraction protocol herein proposed, we applied PCR-based fingerprinting methods starting from the DNA extracted from three EVOO samples of unknown composition. In particular, genomic regions harboring nine simple sequence repeats (SSRs) and eight genotyping-by-sequencing-derived single nucleotide polymorphism (SNP) markers were amplified for authentication and traceability of the three EVOO samples. The whole investigation strategy herein described might favor producers in terms of higher revenues and consumers in terms of price transparency and food safety.

Keywords: DNA extraction protocol; traceability; authentication; genetic tagging; SSRs; SNPs

---

1. Introduction

Olive (*Olea europaea* L. subsp. *europaea* var. *europaea*) is a species that originates from regions surrounding the Mediterranean Sea [1] and characterizes the landscape of the entire Mediterranean area [2]. The olive tree is mainly cultivated for oil production. More than 80% of olive oil production comes from the European Union, with Spain being the largest olive oil producer, followed by Italy and Greece (marketing years 2017/2018) [3]. However, a substantial decrease in olive oil production was observed in Italy in 2018, and it is expected to further decrease in 2019 [4] because of adverse weather conditions and phytosanitary issues that have plagued Italy in the last few years [5,6].

Extra virgin olive oil (EVOO) has elevated commercial value due to its health appeal, desirable characteristics and quantitatively limited production. As a high-value product, a certification label added to a bottle of EVOO allows food fraud to be prevented and consumers to trace oil "from tree to table". The European Commission Regulation 1151/2012 disciplines quality schemes for agricultural products and defines guidelines on the labeling of foodstuffs.

Italy is the first ranked country in terms of the number of protected designation of origin (PDO), protected geographical indication (PGI) and traditional specialty guaranteed (TSG) certifications. As for EVOOs, these include 42 PDO and 4 PGI [7]. Given the importance of quality labeling, all PDO/PGI products have to adhere to specific reference standards. In the case of EVOOs, this means an array of rules on the use of specific olive varieties, the area and practices of cultivation, and the entire production chain from harvest to bottling. Consequently, EVOO traceability is a hot topic since it prevents intentional adulteration accomplished through the deliberate addition of low-cost edible vegetable oils of uncertain origin. As EVOOs on the market might consist of a blend of olive varieties or fraudulently sometimes even of a mixture of oils from different botanical species, varietal identification in olive oils is crucial and has attracted a growing interest in the last decade.

Several approaches based on chemical analysis have been proposed for EVOO traceability [8–12], which are useful to discover some types of macroscopic adulteration, such as the addition of oil from other species. However, they present some limits due to the difficulty in identifying the olive varieties used for oil production. Instead, DNA-based methods are more accurate and reliable than chemical methods; moreover, they are not influenced by environmental conditions [13]. Among the genetic markers used for cultivar identification and varietal traceability in processed foods, simple sequence repeats (SSRs) and single nucleotide polymorphisms (SNPs) are the most suitable [14–16]. Indeed, the use of molecular markers for olive oil traceability is widely documented [17–21].

To have a reliable DNA extraction protocol from EVOOs is the key for a successful molecular fingerprinting. A critical aspect to take into consideration concerns the recovery of an adequate amount of DNA from filtrated oils. DNA is usually degraded and present in a very low amount when a food matrix is used for its extraction [22,23]. So far, diverse methods have been developed to obtain high quality DNA from cellular sediment as well as from the oily and the aqueous phases of filtered EVOOs [24]. All of them share three fundamental steps: (i) cellular lysis; (ii) DNA recovery and (iii) DNA purification. Although each method differs from the other in terms of quality and amount of recovered DNA, all lack information on the effectiveness of the protocol in industrial/legal applications. Moreover, the ideal DNA extraction method should be reproducible, simple, relatively cheap and capable of recovering stable and PCR inhibitor-free DNA from heavily processed plant oil, both filtered and unfiltered. Although different DNA-based methods have been applied for EVOO authentication and traceability, all of them suffer the difficulty of standardizing the DNA extraction protocol as well as the subsequent amplification and genotyping steps.

Within this motivating context, our work presents a case study aiming at set up a timely, low-cost, reproducible and effective DNA extraction protocol from three filtered commercial EVOOs, which allows an adequate amount of DNA to be recovered. Different authors showed the effectiveness of extraction buffer containing CTAB and/or hexane in DNA isolation from different vegetable oils [23–28]. Based on these results, we decided to compare the n-hexane protocol by Consolandi et al. [29] with three CTAB-based methods by Muzzalupo and Perri [30], Busconi et al. [31], and Spadoni et al. [32]. CTAB residual could affect the downstream enzymatic reactions; therefore the detection of an effective and reproducible protocol for DNA extraction and amplification is of crucial importance. Furthermore, as reported by Schrader et al., 2012 [33], different salts (e.g., sodium chloride or potassium chloride), detergents or organic molecules (ethylene-diamine-tetra-acetic acid, sarkosyl, ethanol, isopropyl alcohol or phenol) necessary for efficient cell lysis or for the isolation of pure nucleic acids, might also cause PCR inhibition at certain concentrations. A secondary but no less important purpose was to verify the effectiveness of the DNA extraction protocol herein proposed and to assess PCR-based fingerprinting methods for olive oil authentication and traceability. To this end, we analyzed the DNA polymorphisms of nine SSRs and the presence/absence of eight SNP markers for genetic tagging of Italian EVOOs.

2. Materials and Methods

2.1. Oil Samples and DNA Extraction

Analyses were performed on three EVOO samples (hereinafter referred to as OL1, OL2, and OL3), provided by an Italian mill, which underwent two filtration steps in order to extend their shelf-life. The first step was carried out with two filters made of cellulose and diatomaceous flours; the second filtration step was performed using a polishing paper. The only available information on those samples was that they were bottled in Italy. Three bottles of one liter each were collected and stored at 4 °C before the analysis.

We tested four protocols (hereinafter referred to as p1, p2, p3, and p4) for DNA isolation:

- p1   Spadoni et al., 2019 [32];
- p2   Muzzalupo et al., 2002 [30];
- p3   Busconi et al., 2003 [31];
- p4   Consolandi et al., 2008 [29].

After evaluating those protocols, we modified p4 as follows:

1. One mL per sample was vigorously mixed for 5 min and transferred into a 2 mL microcentrifuge tube, then it was shaken with 500 μL of hexane for 5 min and centrifuged (15 min, 4 °C, 12 krpm).
2. The oily supernatant was transferred to a fresh 2 mL tube and the pellet was mixed with 400 μL of lysis buffer (50 mM Tris pH 8.5, 1 mM EDTA, 0.5% *v/v* Tween 20).
3. Samples were incubated at 48 °C for 1 h and then centrifuged (15 min, 4 °C, 12 krpm).
4. The two aqueous phases from the oil/buffer mix and the pellet were transferred into separate 1.5 mL microcentrifuge tubes, being careful to avoid contact with the interface between oil and buffer.
5. The pellet derived from treatment of the aqueous phase at point 4 was discarded.
6. The three phases were transferred into separate 1.5 mL microcentrifuge tubes and the samples were gently mixed with 500 μL of cold isopropanol, held 30 min at −80 °C, and centrifuged again (30 min, 4 °C, 12 krpm).
7. The pellet was washed by adding 500 μL of absolute ethanol, held 30 min at −80 °C and centrifuged (30 min, 4 °C, 14 krpm).
8. The pellet was washed once more by adding 500 μL of 70% ethanol and centrifuged (30 min, 4 °C, 14 krpm).
9. After removing the supernatant, the pellet was dried in vacuum pump for 30 min and finally resuspended in 50 μL of 1× TE buffer pH 8 (10 mM Tris, 1 mM EDTA).

2.2. SSRs and SNPs Assay

A set of nine SSR primer pairs [34–36] was used to assess the genetic fingerprinting of the three oil samples. Of these, seven (i.e., DCA03, DCA05, DCA09, DCA18, GAPU71B, GAPU101, EMO90) were selected among those available in literature [37] due to their high reproducibility, power of discrimination and number of amplified loci/alleles. The remaining two (i.e., DCA04 and DCA15) were included as they have been proven to be suitable for PCR-based fingerprinting methods in olive [38,39]. PCR was carried out following the procedure by di Rienzo et al. [40]. Sequences and annealing temperatures of each primer are reported in Table S1.

Amplification products were separated by capillary electrophoresis on the ABI PRISM 3100 Avant Genetic Analyzer sequencer (Applied Biosystems, Foster City, CA, USA) using a mix containing 2 μL PCR reaction, 13,5 μL Hi-Di™ Formamide and 0.5 μL GeneScan™ 600 LIZ™ Size Standard (Applied

Biosystem, Foster City, CA, USA). Electropherograms were analyzed by the GeneMapper Software v3.7 (Applied Biosystems, Foster City, CA, USA).

A large panel of 470 Mediterranean olive accessions (from the private database developed by the Department of Soil, Plant and Food Sciences, University of Bari Aldo Moro) including Italian (396), Tunisian (25), Syrian (23), and Algerian (26) accessions, was used to assess genetic similarity among olive accessions and the three commercial EVOOs.

Genetic similarity between the three EVOO samples and the 470 olive accessions was calculated through principal coordinates analysis (PCoA) using the GenAlEx software version 6.5 (http://biology-assets.anu.edu.au/GenAlEx/Download.html). Based on PcoA, accessions with SSR-based genetic profiles more similar to the three samples under study were selected for the construction of a tree through the un-weighted neighbor-joining method, running 1000 bootstrap replicates, using DARWIN v 6.0.010 (http://darwin.cirad.fr) [41]. The resulting tree was visualized using FigTree 2016-10-04-v1.4.3 (http://tree.bio.ed.ac.uk/software/figtree/).

A total of six Italian olive cultivars (Frantoio, Taggiasca, Pendolino, Crastu, Leccino, and Ogliarola barese), representative of the main PDO and PGI certifications (http://www.agraria.org/prodottitipici/oliodioliva.htm), were selected for SNP analysis. We took in consideration the SNP profile from a pre-existing catalogue generated by genotyping-by-sequencing [42]. We selected eight SNPs that satisfied two requirements: (i) each SNP should be polymorphic among the six cultivars; and (ii) the allelic profiles detected using eight SNPs should be specific for each cultivar. Eight primer pairs were designed in SNP flanking sequences.

PCR was performed in a final volume of 20 µL containing 50 ng of DNA, 1X Phusion HF Buffer, 0.2 mM dNTPs Mix, 0.4 µM primer mix and 0.5 units of Phusion high-fidelity DNA polymerase (Thermo Fisher Scientific, Waltham, MA, USA). The cycling program was: 2 min of initial denaturation at 98 °C, 35 cycles of denaturation at 98 °C for 10 sec, annealing for 20 sec, extension at 72 °C for 30 sec and final extension at 72 °C for 5 min. Annealing temperatures ranged from 64 to 68 °C. PCR products were analyzed by 2% agarose gel electrophoresis to verify the specificity of the amplification reaction. Each amplicon was then purified through ethanol precipitation and quantified using the NanoDropTM ND2000C (Thermo Fisher Scientific, Waltham, MA, USA).

PCR products obtained from DNA extracted from oils required an additional step of purification, which was performed using the Agencourt AMPure XP - PCR Purification Kit (Beckman Coulter, USA). Sequencing reaction was prepared using the BigDye™ Terminator v3.1 Cycle Sequencing Kit following the manufacturer's instructions and the sequencing was performed by the ABI PRISM 3100 Avant Genetic Analyzer (Applied Biosystems, Foster City, CA, USA).

Finally, raw data were analyzed by the Sequencing Analysis Software v6.0 and the Sequence Scanner Software v2.0.

3. Results

3.1. Set Up of An Effective Protocol for DNA Extraction from Filtered Oils and EVOO Traceability

As a first step, the four protocols (i.e., p1–p4) were compared each other to assess their effectiveness in extracting DNA from the three filtered EVOO samples. This was done to identify the most suitable one to be modified if necessary (Table 1). Although p1 was developed to extract DNA from olive leaves, we have also chosen to assess its effectiveness because of its rapidity and flexibility.

All protocols share an initial step of sample incubation at 60–65 °C with a standard extraction buffer consisting of Tris-HCl and EDTA. All protocols, except p4, are based on the use of CTAB, while p1 includes the addition of the anionic detergent sodium dodecyl sulfate (Table 1). Afterwards, organic solvents such as hexane, dichloromethane, octanol, and phenol were used for the dissolution of the plasma membrane. DNA precipitation was obtained with ethanol or isopropanol.

Table 1. Comparison among the DNA extraction methods used in this study. Extraction time, starting amount of plant tissue/oil (* as recommended by the authors), and reagents used are indicated.

	P1	P2	P3	P4	P5
	Spadoni et al., 2019 [32]	Muzzalupo et al., 2002 [30]	Busconi et al., 2003 [31]	Consolandi et al., 2008 [29]	Consolandi et al., 2008 modified [29]
Extraction time (hours)	~6 h	~30 h	~6 h	~30 h	~4 h
Sample/tissue and amount *	2 g leaves	10 mL not filtered, clear oil	50 mL not filtered, clear oil	2 mL not filtered, clear oil	1 mL filtered, clear oil
Starting centrifugation step	No	Yes	Yes	No	No
Main solvent	CTAB, phenol, chloroform	CTAB, dichloromethane, chloroform	CTAB, octanol, chloroform	Hexane, chloroform	Hexane, chloroform
Liquid nitrogen	No	No	Yes	No	No
Proteinase K	No	No	No	Yes	No
Pronase	No	Yes	No	No	No
SDS	Yes	No	No	No	No
RNAse treatment	No	Yes	Yes	No	No
β-mercaptoethanol	Yes	Yes	Yes	No	No
Extraction time (hours)	~6 h	~30 h	~6 h	~30 h	~4 h

Each protocol was tested following authors' indications. After extraction, DNA was checked for quantity and quality. In all cases, DNA was highly degraded and its amount ranged from 0.5 ng/μL (p1) to 100 ng/μL (p4).

P1 was discarded because of the low amount of recovered DNA (<2 ng/μL). P2 and p3 were both based on the use of CTAB. Although both protocols start from a volume of oil higher than 10 mL, the amount of recovered DNA was lower than 10 ng/μL. Moreover, the use of CTAB, which is an inhibitor of enzymatic reactions, requires several washing steps to remove its traces. By contrast, the extraction solvent used in p4 is the hexane and this protocol resulted in a fair amount of DNA, ranging from 20 to 100 ng/μL, though starting from 2 mL of EVOO. The only drawback is that p4 was the most time-consuming. Therefore, we decided to improve p4 in order to (i) reduce the time required for the DNA extraction process; (ii) decrease the starting amount of oil; and (iii) limit the number of reagents and equipment needed. All these improvements are crucial for industrial and legal application of the method.

The protocol herein proposed (labeled as p5 in Table 1) was obtained through the optimization of p4 with some key modifications. Firstly, a reduction of the starting material (to 1 mL) was applied; this change allowed the starting centrifugation step included in p4 to be eliminated and replaced by vigorous stirring for a few minutes in a ThermoMixer. Furthermore, the sample digestion step with Proteinase K was ruled out, thus reducing to 1 h the incubation time of sample after the addition of the lysis buffer. Finally, two steps of DNA precipitation were performed. The first one was carried out in isopropanol with incubation at −80 °C for 30 min instead of −20 °C overnight. The second step was performed in absolute ethanol with incubation at −80 °C for 30 min. Nucleic acid quality evaluation was performed by 1% of agarose gel (Figure S1) and showed a high degradation of genomic DNA. The quantity values, checked through Nanodrop, ranged between 53.8 ng/μL and 442.7 ng/μL for DNA extracted from oily and aqueous phases and between 0.4 ng/μL and 9.5 ng/μL for DNA extracted from the pellet phase.

3.2. EVOO Genotyping and Traceability

To test the suitability and effectiveness of the protocol herein proposed and set up a reproducible genetic tagging and authentication method, genomic regions harboring SSR and SNP markers were amplified starting from the DNA extracted from the three oil samples. For each sample, three independent DNA extraction replicates were performed, in order to assess the repeatability of the technique.

First of all, in order to evaluate the suitability of the extracted DNA to PCR amplification, we performed loci specific PCR amplifications using the microsatellite markers DCA03 and DCA18, which have proven to be highly reproducible based on our experience [43]. PCR products were obtained using DNA extracted from oily, aqueous and pellet phases (Figure S2), showing 100% repeatability (10 technical replicates). All samples provided a clear and distinguishable amplification pattern (Figure 1).

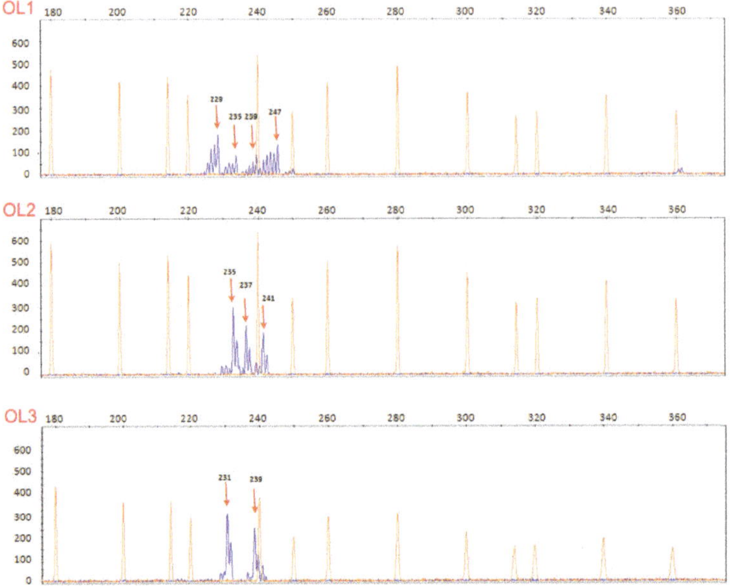

Figure 1. Capillary electropherograms showing the dinucleotide DCA03 amplification pattern for OL1, OL2 and OL3. Allele size (in base pairs) is indicated in correspondence of the main peak.

For practical reasons, PCR runs for all SSR markers were performed on DNA extracted from aqueous phase only. The three EVOO samples showed monomorphic profiles for DCA04, DCA09, EMO90, and GAPU71B, while different allele combinations were recorded for the remaining markers in each sample (Figure 2a). The most informative markers were DCA03 and DCA18, for OL1 and OL3, respectively, as each of them discriminated among 10 different allele combinations. The total number of alleles was 22 for all oils, even the allelic variants were different. We compared the pattern of SSR allele sizes with a database of SSR-based allele frequencies of 470 records including Italian, Tunisian, Algerian, Syrian, and native Apulian varieties, developed at the Department of Soil, Plant and Food Sciences, University of Bari "Aldo Moro" (Figure 2b).

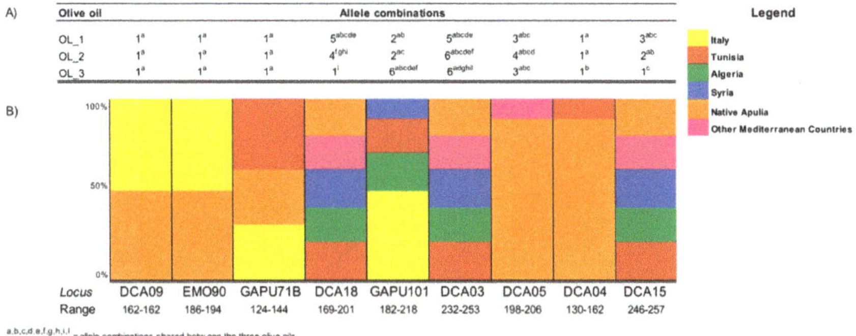

Figure 2. Allelic patterns detected in the three olive oils using nine simple sequence repeat (SSR) markers. (**A**) The number and composition of allelic combinations were reported for each extra virgin olive oil (EVOO) sample. (**B**) Bar plot representing the proportion of allelic combination for each of the nine SSR *loci*.

We included a set of 38 olive accessions collected in Apulia, which is the major oil-producing region in the Southern Italy [39]. Indeed, many common national varieties used in the EVOO production (i.e., Coratina, Cima di Bitonto, Ogliarola barese, etc.) derived from Apulia.

To explore genetic similarities among the three EVOO samples and the 470 olive accessions, principal coordinates analysis (PCoA) was performed (Figure 3).

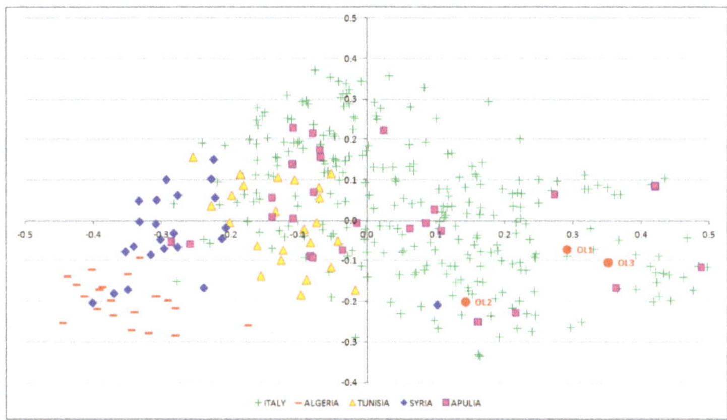

Figure 3. Principal coordinates analysis (PCoA) performed using the three EVOO samples and the panel of 470 Mediterranean olive accessions. OL1, OL2 and OL3 are marked in red.

The three samples resulted to be different but very close each other, suggesting the use of common and related varieties in EVOO production. Moreover, all the samples were placed in a group that includes the main Italian olive varieties. A subset of 147 accessions that lie in the fourth quadrant of the PCoA plot together with the three EVOO samples was filtered out and used to construct a tree through the un-weighted neighboring joining method (Figure 4). The dendrogram disclosed that the three EVOOs were grouped in the same cluster; however, OL1 and OL3 are more similar to each other. Finally, the clade in which the three EVOO samples fell includes some of the most widespread Italian varieties used for oil production, such as Frantoio, Cima di Bitonto and other many olive accessions originated from Apulia.

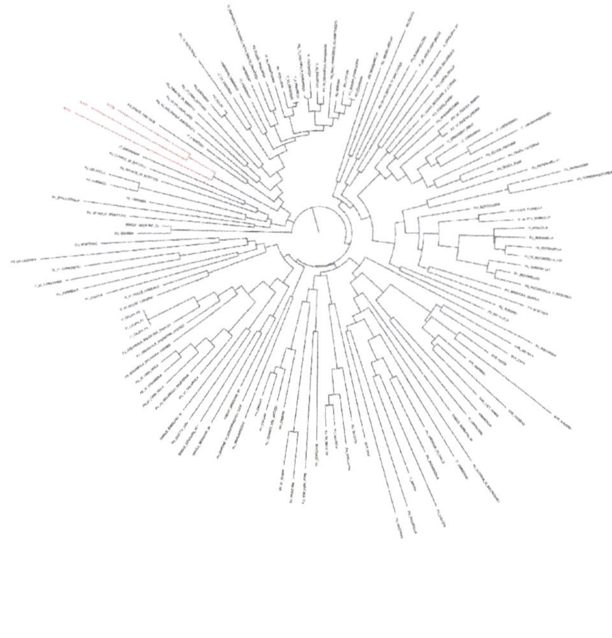

Figure 4. Genetic similarity between the three EVOO samples and the subset of 147 olive cultivars selected on the bases of PCoA results. OL1, OL2 and OL3 are marked in red.

Eight SNPs, which discriminate against six varieties, namely Frantoio, Taggiasca, Pendolino, Crastu, Leccino, and Ogliarola barese, were selected and the DNA regions harboring those SNPs subjected to PCR amplification and Sanger sequencing. PCR experiments were successful in all the samples (Table 2).

Table 2. List of single nucleotide polymorphisms (SNPs) validated in the selected varieties as well as in the three EVOO samples. n.a. = not available.

Variety	# SNP							
	1	2	3	4	5	6	7	8
Frantoio	CC	GG	TC	GG	AT	GG	GG	AT
Taggiasca	CC	GT	TC	GG	AT	GC	GG	AT
Pendolino	CC	GG	TT	AG	TT	GC	AA	AA
Crastu	TT	TT	TT	AG	TT	GG	AA	AA
Leccino	CC	GG	TC	GG	TT	GG	AA	AT
Ogliarola barese	CC	GT	n.a.	GG	AT	CC	GG	AT
OL1	CC	GT	TT	GG	TT	GG	AA	AT
OL2	CC	GT	TT	n.a.	TT	GG	AA	AT
OL3	CC	GT	TT	GG	TT	GG	GG	AT

The presence of non-nucleic acid contaminants and inhibitors in the amplified DNA involved an additional purification step, following which sequencing reactions definitely improved as shown by the severe reduction of the background noise (Figure S3).

All samples were characterized by the same alleles at seven SNP sites (from SNP #1 to #7) with the exception of OL3, which showed a different allele for SNP #7 (Figure 5). As for SNP #8, differences in the allelic profile were found across all samples (Figure 6).

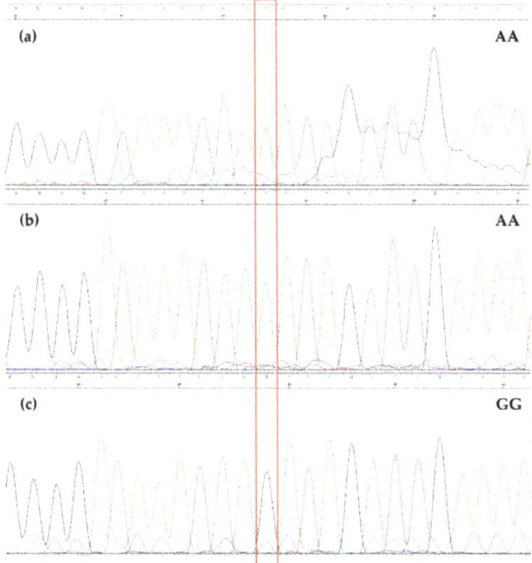

Figure 5. Electropherogram obtained from Sanger sequencing of the region containing SNP #7 in OL1 (**a**), OL2 (**b**), and OL3 (**c**). A red box highlights the SNP.

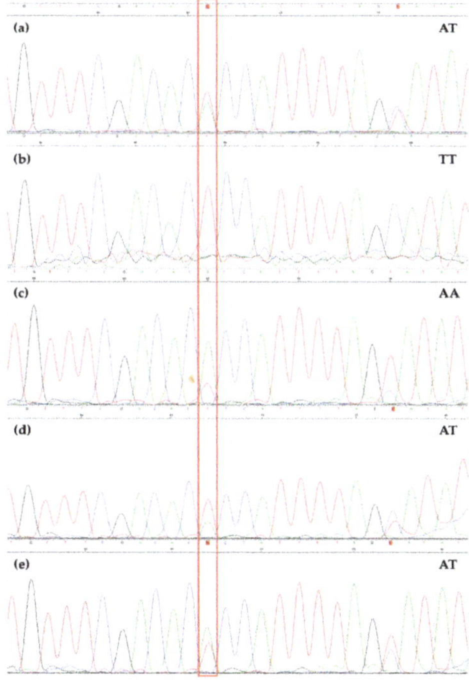

Figure 6. Electropherogram obtained from Sanger sequencing of the region containing SNP #8 in Ogliarola barese (**a**), Pendolino (**b**), OL1 (**c**), OL2 (**d**), and OL3 (**e**). A red box highlights the SNP.

4. Discussion

When olive drupes are ground, the cellular compartmentalization is destroyed and everything is mixed in the aqueous solution (i.e., waste water) where most of the DNA and proteins are. Olive oil is the water-insoluble fraction mainly lacking in DNA and proteins, but enriched in organic compounds that interfere with DNA polymerase reactions.

Genetic traceability of EVOOs requires DNA to be extracted from commercial olive oils, which are all subjected to a filtration process in order to satisfy consumers' demand (i.e., clear oil).

To identify an effective and robust DNA extraction protocol from filtered commercial oils, four protocols were tested and compared each other (Table 1). However, none of these proved to be highly reproducible as the extraction of DNA from a complex matrix to be subjected to PCR is not a trivial task [15,44,45]. Following the comparison, only p4 turned out to be the most suitable in terms of DNA yield, but it did not guarantee repeatable results with regards to DNA amplification. Therefore, we improved p4 with key modifications in order to enhance the quality and quantity of the extracted DNA while reducing the starting material, reagents, costs and time of extraction. The protocol herein proposed (labeled as p5 in Table 1) was applied on three EVOO samples of unknown composition. Nanodrop measurements were satisfactory for the DNA extracted from the oil and aqueous phases with values ranging from 442.7 ng/µL to 53.8 ng/µL. By contrast, DNA concentration from the pellet phase ranged from 9.5 ng/µL to 0.4 ng/µL. Taking into account the nature of the starting material, we have obtained appreciable quantity of DNA from all three extraction phases.

The effectiveness of p5 was assessed by applying PCR-based fingerprinting methods for EVOO authentication and traceability. DNA amplifications of nine SSR loci were successful for all the three EVOO samples.

The presence of microsatellite multi-allelic combinations (Figure 2) coupled with results from PCoA and genetic similarity analysis (Figures 3 and 4), both describing the relationships between the three EVOO samples and a large panel of accessions cultivated in the Mediterranean area, allowed us to assert that most likely the three EVOO samples are a blend of Italian varieties.

If SNP markers are used for EVOO authentication and traceability, we suggest using an additional purification step of PCR products. Indeed, phenolic compounds and residual polysaccharides could inhibit the enzymatic reactions providing irregular PCR amplifications [46]. This inhibition is negligible in routine PCRs, such as those for SSR amplification, but it is exacerbated in highly sensitive reactions, such as Sanger sequencing. Then, we applied an additional PCR purification step based on the use of magnetic beads after ethanol precipitation, which allowed us to gain a significant improvement in sequencing results (Figure S3). Previous works performed magnetic bead-based purification before PCR [47,48]; however, this procedure leads to a massive loss of DNA. As the amount of DNA extracted from filtrated oil was generally low, we decided to carry out this step after PCR amplification.

Advances in technologies for DNA profiling have had a huge impact in combating fraud and food adulteration as well as in improving authentication and traceability of food products. The screening of multiple markers with next generation sequencing (NGS) technology has been widely applied to the olive sector, and in particular the high-resolution DNA melting technology (HRM) has been successfully used for olive oil authentication [49,50]. HRM analysis using gene-based SNP markers represents an important tool for cultivar identification. However, gDNA extraction from olive oil matrix is still a critical step, which needs improvement [51]. For that reason, we chose to amplify the DNA of the three EVOOs with SNP-based markers and then perform sequencing by Sanger method. SNP-based molecular profiles revealed high similarity among the three EVOO samples, thus indicating that they shared some varieties, as observed before. By comparing these profiles with those obtained for six Italian olive cultivars, we observed that OL1 and OL2 were very similar to Leccino, while OL3 is more similar to Frantoio and Taggiasca (Table 2).

Despite no information about varietal composition of the three EVOO samples under investigation was available, we were able to assert that most likely the three samples are blends of Italian varieties and generate a list of possible varieties present in each blend.

The EU commission regulation 182/2009 governs the whole production process of EVOOs. The origin of olives and/or the addition of oils coming from different countries must be indicated on the label. Molecular traceability is an important tool to investigate the varietal composition of olive oils, in order to prevent fraud due to the use of oils coming from extra-EU countries not explicitly indicated on the label.

Contrary to what happens for the wine market, in which there are few international varieties (i.e., Cabernet, and Merlot) that are cultivated widespread in the world, for the oil market there are cultivars strongly adapted and cultivated in each territory (i.e., Arbequinia and Arbosana for Spain, Chemlali for Tunisia, Coratina and Frantoio for Italy). In this context, the identification of varietal composition could be useful to better define the origin of the oil.

Nevertheless, the comprehensive identification of all the varieties present in a blend is extremely tricky. The identification of singleton SNPs and/or haplotype blocks via NGS will allow the rapid detection of cultivars present in a blend in the near future.

5. Conclusions

We developed a speedy and highly reproducible protocol able to recover whole genomic DNA from filtrated EVOOs. Then, we applied PCR-based methods for genetic tagging of filtered EVOOs of unknown origin in order to verify the effectiveness of the proposed DNA extraction protocol. The whole investigation strategy may be suitable for industrial/legal applications, as olive oil frauds are more and more rampant. Deliberate or unintentional adulteration of olive oil could seriously damage producers/providers in term of earnings and reputation and could definitely break the producer–consumer relationship. On the other hand, consumers are increasingly demanding as their preferences are moving towards the highest quality products. There is a vast range of olive oil on the market and consumers would like to have awareness of the entire olive oil supply chain, "from tree to table". DNA-based methods for olive oil authentication and traceability are beginning to be routinely applied, and although they are still challenging due to the complexity of fraudulent practices, they can reassure consumers in terms of price transparency and food safety.

Supplementary Materials: The following are available online at http://www.mdpi.com/2304-8158/8/10/462/s1, Table S1: List of SSR markers used for EVOO traceability. Primer sequences and annealing temperature are also provided. Figure S1: Electrophoresis on 1% agarose gel of DNA extracted from samples OL1 (1), OL2 (2), and OL3 (3) in three phases (pellet, oil, and aqueous). Figure S2: Electrophoresis on 1% agarose gel of amplification product on samples OL1, OL2, and OL3 in three phases (oil, aqueous, and pellet) with GAPU71B. Figure S3: Electropherogram obtained from Sanger sequencing of a genomic region amplified in OL1 before (**a**) and after (**b**) purification step.

Author Contributions: Conceptualization, L.P., F.T., R.S., C.M. and V.F. (Valentina Fanelli); Data curation, M.A.S.; Formal analysis, L.P., M.A.S., F.T. and S.G. (Susanna Gadaleta); Investigation, L.P., R.S., C.M. and V.F. (Valentina Fanelli); Methodology, L.P., M.A.S., S.G. (Susanna Gadaleta), C.D.G. and A.P.; Resources, S.G. (Stefania Girone); Software, F.T., N.D.A., S.G. (Stefania Girone), V.F. (Vincenzo Fucilli) and V.F. (Valentina Fanelli); Supervision, C.M. and V.F. (Valentina Fanelli); Validation, N.D.A.; Visualization, M.A.S. and A.P.; Writing—original draft, L.P., M.A.S., F.T., C.M. and V.F. (Valentina Fanelli); Writing—review & editing, F.T., N.D.A., C.M. and V.F. (Valentina Fanelli).

Funding: This research was supported by the regional government of Apulia within the PROGRAMMA SVILUPPO RURALE FEASR 2014–2020 Asse II "Miglioramento dell'Ambiente e dello Spazio Rurale" Misura 10.2.1 "Progetti per la conservazione e valorizzazione delle risorse genetiche in agricoltura"—trascinamento della Misura 214 Az. 4 sub azione (a) del PSR 2007–2013 Progetti integrati per la biodiversità—Progetto Re.Ger.O.P. "Recupero del Germoplasma Olivicolo Pugliese" Progetto di continuità.

Conflicts of Interest: The authors declare no conflict of interest.

References

1. Besnard, G.; Terral, J.F.; Cornille, A. On the origins and domestication of the olive: A review and perspectives. *Ann. Bot.* **2018**, *121*, 385–403. [CrossRef]
2. Sardaro, R.; Bozzo, F.; Petrontino, A.; Fucilli, V. Community preferences in support of a conservation programme for olive landraces in the Mediterranean area. *Acta Hort.* **2018**, *1199*, 183–188. [CrossRef]

3. International Olive Council. Available online: http://www.internationaloliveoil.org/ (accessed on 26 August 2019).

4. Ismea–Istituto di Servizi per il Mercato Agricolo Alimentare. Available online: http://www.ismea.it/istituto-di-servizi-per-il-mercato-agricolo-alimentare (accessed on 26 August 2019).

5. Martelli, G.P.; Boscia, D.; Porcelli, F.; Saponari, M. The olive quick decline syndrome in south-east Italy: A threatening phytosanitary emergency. *Eur. J. Plant Pathol.* **2016**, *144*, 235–243. [CrossRef]

6. Tanasijevic, L.; Todorovic, M.; Pereira, L.S.; Pizzigalli, C.; Lionello, P. Impacts of climate change on olive crop evapotranspiration and irrigation requirements in the Mediterranean region. *Agric. Water Manag.* **2014**, *144*, 54–68. [CrossRef]

7. Istat. Available online: https://www.istat.it/ (accessed on 8 August 2019).

8. Likudis, Z. Olive Oils with Protected Designation of Origin (PDO) and Protected Geographical Indication (PGI). In *Products from Olive Tree*; Boskou, D., Ed.; IntechOpen Publisher: London, UK, 2016; pp. 175–190.

9. Bhandari, M.P.; Carmona, E.N.; Abbatangelo, M.; Sberveglieri, V.; Duina, G.; Malla, R.; Sberveglieri, G. Discrimination of Quality and Geographical Origin of Extra Virgin Olive Oil by S3 Device with Metal Oxides Gas Sensors. *Multidiscip. Digit. Publ. Inst. Proc.* **2018**, *2*, 1061. [CrossRef]

10. Kalaitzis, P.; El-Zein, Z. Olive oil authentication, traceability and adulteration detection using DNA-based approaches. *Lipid Technol.* **2016**, *28*, 10–11. [CrossRef]

11. Meenu, M.; Cai, Q.; Xu, B. A critical review on analytical techniques to detect adulteration of extra virgin olive oil Trends. *Food Sci. Technol.* **2019**, *91*, 391–408. [CrossRef]

12. Xanthopoulou, A.; Ganopoulos, I.; Bosmali, I.; Tsaftaris, A.; Madesis, P. DNA fingerprinting as a novel tool for olive and olive oil authentication, traceability, and detection of functional compounds. In *Olives and Olive Oil as Functional Foods: Bioactivity, Chemistry and Processing*; Shahidi, F., Kiritsakis, A., Eds.; John Wiley & Sons Ltd.: Hoboken, NJ, USA, 2017; pp. 587–601.

13. Bazakos, C.; Spaniolas, S.; Kalaitzis, P. DNA-Based Approaches for Traceability and Authentication of Olive Oil. In *Products from Olive Tree*; Boskou, D., Ed.; IntechOpen Publisher: London, UK, 2016; pp. 115–133.

14. Espiñeira, M.; Santaclara, F. *Advances in Food Traceability Techniques and Technologies*; Woodhead Publishing: Cambridge, UK, 2016.

15. Di Rienzo, V.; Miazzi, M.M.; Fanelli, V.; Savino, V.; Pollastro, S.; Colucci, F.; Miccolupo, A.; Blanco, A.; Pasqualone, A.; Montemurro, C. An enhanced analytical procedure to discover table grape DNA adulteration in industrial musts. *Food Control* **2016**, *60*, 124–130. [CrossRef]

16. Pasqualone, A.; di Rienzo, V.; Nasti, R.; Blanco, A.; Gomes, T.; Montemurro, C. Traceability of Italian Protected Designation of Origin (PDO) Table Olives by Means of Microsatellite Molecular Markers. *J. Agr. Food Chem.* **2013**, *61*, 3068–3073. [CrossRef] [PubMed]

17. Uncu, A.T.; Frary, A.; Doganlar, S. Cultivar Origin and Admixture Detection in Turkish Olive Oils by SNP-Based CAPS Assays. *J. Agric. Food Chem.* **2015**, *63*, 2284–2295. [CrossRef] [PubMed]

18. Pasqualone, A.; Montemurro, C.; di Rienzo, V.; Summo, C.; Paradiso, V.M.; Caponio, F. Evolution and perspectives of cultivar identification and traceability from tree to oil and table olives by means of DNA markers. *J. Sci. Food Agric.* **2016**, *96*, 3642–3657. [CrossRef] [PubMed]

19. Montemurro, C.; Miazzi, M.M.; Pasqualone, A.; Fanelli, V.; Sabetta, W.; di Rienzo, V. Traceability of PDO Olive Oil "Terra di Bari" Using High Resolution Melting. *J. Chem.* **2015**, *205*, 1–7. [CrossRef]

20. Raieta, K.; Muccillo, L.; Colantuoni, V. A novel reliable method of DNA extraction from olive oil suitable for molecular traceability. *Food Chem.* **2015**, *172*, 596–602. [CrossRef] [PubMed]

21. Ben-Ayed, R.; Grati-Kamoun, N.; Sans-Grout, C.; Moreau, F.; Rebai, A. Characterization and authenticity of virgin olive oil (*Olea europaea* L.) cultivars by microsatellite markers. *Eur. Food Res. Technol.* **2012**, *234*, 263–271. [CrossRef]

22. Hrnčírová, Z.; Bergerová, E.; Siekel, P. Effects of technological treatment on DNA degradation in selected food matrices of plant origin. *J. Food Nutr. Res.* **2008**, *47*, 23–28.

23. De la Torre, F.; Bautista, R.; Cánovas, F.M.; Claros, G. Isolation of DNA from olive oil and oil sediments: Application in oil fingerprinting. *Food Agric. Environ.* **2004**, *2*, 84–89.

24. Montemurro, C.; Sabetta, W.; di Rienzo, V.; Pasqualone, A. Advances in DNA Extraction for Detecting Extra-Virgin Olive Oil Adulteration. In *Authentication and Detection of Adulteration of Olive Oil*; Kontominas, M., Ed.; Nova Science: New York, NY, USA, 2018; pp. 315–328.

25. Ramos-Gómez, S.; Busto, M.D.; Perez-Mateos, M.; Ortega, N. Development of a method to recovery and amplification DNA by real-time PCR from commercial vegetable oils. *Food Chem.* **2014**, *158*, 374–383. [CrossRef]

26. Martins-Lopes, P.; Gomes, S.; Santos, E.; Guedes-Pinto, H. DNA Markers for Portuguese Olive Oil Fingerprinting. *J. Agr. Food Chem.* **2008**, *56*, 11786–11791. [CrossRef]

27. Giménez, M.J.; Pistón, F.; Martín, A.; Atienza, S.G. Application of real-time PCR on the development of molecular markers and to evaluate critical aspects for olive oil authenticatio. *Food Chem.* **2010**, *118*, 482–487.

28. Gomes, S.; Breia, R.; Carvalho, T.; Carnide, V.; Martins-Lopes, P. Microsatellite High-Resolution Melting (SSR-HRM) to Track Olive Genotypes: From Field to Olive Oil. *J. Food Sci.* **2018**, *8*, 2415–2423. [CrossRef]

29. Consolandi, C.; Palmieri, L.; Severgnini, M.; Maestri, E.; Marmiroli, N.; Agrimonti, C.; Baldoni, L.; Donini, P.; De Bellis, G.; Castiglioni, B. A procedure for olive oil traceability and authenticity: DNA extraction, multiplex PCR and LDR-universal array analysis. *Eur. Food Res. Technol.* **2008**, *227*, 1429–1438. [CrossRef]

30. Muzzalupo, I.; Perri, E. Recovery and characterization of DNA from virgin olive oil. *Eur. Food Res. Technol.* **2002**, *214*, 528–531. [CrossRef]

31. Busconi, M.; Foroni, C.; Corradi, M.; Bongiorni, C.; Cattapan, F.; Fogher, C. DNA extraction from olive oil and its use in the identification of the production cultivar. *Food Chem.* **2003**, *83*, 127–134. [CrossRef]

32. Spadoni, A.; Sion, S.; Gadaleta, S.; Savoia, M.A.; Piarulli, L.; Fanelli, V.; di Rienzo, V.; Taranto, F.; Miazzi, M.M.; Montemurro, C.; et al. A simple and rapid method for genomic DNA extraction and microsatellite analysis in tree plants. *J. Agr. Sci. Technol.* **2019**, *21*, 1215–1226.

33. Schrader, C.; Schielke, A.; Ellerbroek, L.; Johne, R. PCR inhibitors–Occurrence, properties and removal. *J. App. Microbiol.* **2012**, *113*, 1014–1026. [CrossRef] [PubMed]

34. Carriero, F.; Fontanazza, G.; Cellini, F.; Giorio, G. Identification of simple sequence repeats (SSRs) in olive (*Olea europaea* L.). *Theor. Appl. Genet.* **2002**, *104*, 301–307. [CrossRef] [PubMed]

35. De la Rosa, R.; James, C.M.; Tobutt, K.R. Isolation and characterization of polymorphic microsatellites in olive (*Olea europaea* L.) and their transferability to other genera in the *Oleaceae*. *Mol. Ecol. Notes* **2002**, *1*, 346–348. [CrossRef]

36. Sefc, K.M.; Lopes, M.S.; Mendonça, D.; Rodrigues Dos Santos, M.; Laimer Da Câmara Machado, M.; Da Câmara Machado, A. Identification of microsatellite loci in olive (*Olea europaea*) and their characterization in Italian and Iberian olive trees. *Mol. Ecol.* **2000**, *9*, 1171–1173. [CrossRef] [PubMed]

37. Baldoni, L.; Cultrera, N.G.; Mariotti, R.; Ricciolini, C.; Arcioni, S.; Vendramin, G.G.; Buonamici, A.; Porceddu, A.; Sarri, V.; Ojeda, M.A.; et al. A consensus list of microsatellite markers for olive genotyping. *Mol. Breed.* **2009**, *24*, 213–231. [CrossRef]

38. Alba, V.; Montemurro, C.; Sabetta, W.; Pasqualone, A.; Blanco, A. SSR-based identification key of cultivars of *Olea europaea* L. diffused in Southern-Italy. *Sci. Hortic.* **2009**, *123*, 11–16. [CrossRef]

39. Sion, S.; Taranto, F.; Montemurro, C.; Mangini, G.; Camposeo, S.; Falco, V.; Gallo, A.; Mita, G.; Debbabi, O.S.; Amar, F.B.; et al. Genetic Characterization of Apulian Olive Germplasm as Potential Source in New Breeding Programs. *Plants* **2019**, *8*, 268. [CrossRef] [PubMed]

40. Di Rienzo, V.; Sion, S.; Taranto, F.; D'Agostino, N.; Montemurro, C.; Fanelli, V.; Sabetta, W.; Boucheffa, S.; Tamendjari, A.; Pasqualone, A.; et al. Genetic flow among olive populations within the Mediterranean basin. *PeerJ* **2018**, *6*, e5260. [CrossRef]

41. Perrier, X.; Flori, A.; Bonnot, F. Data Analysis Methods. In *Genetic Diversity of Cultivated Tropical Plants*; Hamon, P., Seguin, M., Perrier, X., Glaszmann, J.C., Eds.; Science Publishers: Enfield, NH, USA, 2003; pp. 43–76.

42. D'Agostino, N.; Taranto, F.; Camposeo, S.; Mangini, G.; Fanelli, V.; Gadaleta, S.; Miazzi, M.M.; Pavan, S.; di Rienzo, V.; Sabetta, W.; et al. GBS-derived SNP catalogue unveiled wide genetic variability and geographical relationships of Italian olive cultivars. *Sci. Rep.* **2018**, *8*, 15877.

43. Pasqualone, A.; di Rienzo, V.; Miazzi, M.M.; Fanelli, V.; Caponio, F.; Montemurro, C. High resolution melting analysis of DNA microsatellites in olive pastes and virgin olive oils obtained by talc addition. *Eur. J. Lipid Sci. Tech.* **2015**, *117*, 2044–2048. [CrossRef]

44. Pasqualone, A.; Montemurro, C.; Grinn-Gofron, A.; Sonnante, G.; Blanco, A. Detection of soft wheat in semolina and durum wheat bread by analysis of DNA microsatellites. *J. Agr. Food Chem.* **2007**, *55*, 3312–3318. [CrossRef] [PubMed]

45. Pasqualone, A.; Alba, V.; Mangini, G.; Blanco, A.; Montemurro, C. Durum wheat cultivar traceability in PDO Altamura bread by analysis of DNA microsatellites. *Eur. Food Res. Technol.* **2010**, *230*, 723–729. [CrossRef]
46. Agrimonti, C.; Vietina, M.; Pafundo, S.; Marmiroli, N. The use of food genomics to ensure the traceability of olive oil. *Trends Food Sci. Technol.* **2011**, *22*, 237–244. [CrossRef]
47. Breton, C.; Claux, D.; Metton, I.; Skorski, G.; Bervilleä, A. Comparative Study of Methods for DNA Preparation from Olive Oil Samples to Identify Cultivar SSR Alleles in Commercial Oil Samples: Possible Forensic Applications. *J. Agr. Food Chem.* **2004**, *52*, 531–537. [CrossRef]
48. Bracci, T.; Busconi, M.; Fogher, C.; Sebastiani, L. Molecular studies in olive (*Olea europaea* L.): Overview on DNA markers applications and recent advances in genome analysis. *Plant Cell Rep.* **2011**, *30*, 449–462. [CrossRef] [PubMed]
49. Vietina, M.; Agrimonti, C.; Marmiroli, N. Detection of plant oil DNA using high resolution melting (HRM) post PCR analysis: A tool for disclosure of olive oil adulteration. *Food Chem.* **2013**, *14*, 3820–3826. [CrossRef] [PubMed]
50. Ganopoulos, I.; Bazakos, C.; Madesis, P.; Kalaitzis, P.; Tsaftaris, A. Barcode DNA high-resolution melting (Bar-HRM) analysis as a novel close-tubed and accurate tool for olive oil forensic use. *J. Sci. Food Agric.* **2013**, *93*, 2281–2286. [CrossRef] [PubMed]
51. Pereira, L.; Gomes, S.; Barrias, S.; Fernandes, J.R.; Martins-Lopes, P. Applying high-resolution melting (HRM) technology to olive oil and wine authenticity. *Food Res. Int.* **2018**, *103*, 170–181. [CrossRef] [PubMed]

Article

Authentication of the Geographical Origin of Margarines and Fat-Spread Products from Liquid Chromatographic UV-Absorption Fingerprints and Chemometrics

Sanae Bikrani [1], Ana M. Jiménez-Carvelo [2,*], Mounir Nechar [1], M. Gracia Bagur-González [2], Badredine Souhail [1] and Luis Cuadros-Rodríguez [2]

[1] Department of Chemistry, Faculty of Sciences, University of Abdelmalek Essaâdi, Av. Sebta, Mhannech II, 93002 Tetouan, Morocco; sanae24bikrani@gmail.com (S.B.); mnechar@gmail.com (M.N.); bssouhail@yahoo.fr (B.S.)

[2] Department of Analytical Chemistry, Faculty of Sciences, University of Granada, C/ Fuentenueva, s/n, E-18071 Granada, Spain; mgbagur@ugr.es (M.G.B.-G.); lcuadros@ugr.es (L.C.-R.)

* Correspondence: amariajc@ugr.es; Tel.: +34-958-24-0797

Received: 15 October 2019; Accepted: 16 November 2019; Published: 19 November 2019

Abstract: Fat-spread products are a stabilized emulsion of water and vegetable oils. The whole fat content can vary from 10 to 90% (*w/w*). There are different kinds, which are differently named, and their composition depends on the country in which they are produced or marketed. Thus, having analytical solutions to determine geographical origin is required. In this study, some multivariate classification methods are developed and optimised to differentiate fat-spread-related products from different geographical origins (Spain and Morocco), using as an analytical informative signal the instrumental fingerprints, acquired by liquid chromatography coupled with a diode array detector (HPLC-DAD) in both normal and reverse phase modes. No sample treatment was applied, and, prior to chromatographic analysis, only the samples were dissolved in n-hexane. Soft independent modelling of class analogy (SIMCA) and partial least squares-discriminant analysis (PLS-DA) were used as classification methods. In addition, several classification strategies were applied, and performance of the classifications was evaluated applying proper classification metrics. Finally, 100% of samples were correctly classified applying PLS-DA with data collected in reverse phase.

Keywords: liquid chromatography fingerprinting; food authentication; margarines and spreads; multivariate classification

---

1. Introduction

Margarine is a water-in-oil emulsion derived from vegetable and/or animal fats, with a fat content ranging from 80 to 90% (*w/w*) and a milk fat content of no more than 3% (*w/w*) [1,2]. This emulsion remains solid at 20 °C. It was discovered by Hippolyte Mège Mouriés, a French chemist, who patented his product in France and Britain in 1869. The product was described at that time as a blend of the glycerol esters of oleic and margaric acids and was therefore called oleo-margarine. Margaric acid was thought to be heptadecanoic acid (17:0), but currently, it is known that it was really a eutectic mixture of palmitic (16:0) and stearic (18:0) acids [3]. "Fat-spread" is the generic name that is applied to any emulsion with a fat content ranging from 10 to 90%, whereas the term "vegetable shortening" or just "shortening" refers to all fat produced from vegetal oils that are hydrogenated, i.e., they stay semisolid at room temperature. According to this, the fat content of shortening is 100%.

The European legislation currently in force distinguishes the following types, regarding fat content [4]: Margarine, three-quarter fat margarine, and half-fat margarine. The half-fat margarine is also named minarine or halvarine by Codex Alimentarius terminology [2]. On the other hand, Moroccan legislation is less restrictive and does not consider different kinds of margarine concerning fat content. In fact, margarine is defined as "any food fat substance, which is not butter or lard, but resembles butter and it is produced for using as butter" [5]. The milk fat content cannot exceed 10% (*w/w*). In order to simplify matters, and because it would not have major repercussions on the study, the term "margarine/spread" will be used in a broad sense in this paper, regardless of fat content.

For the manufacturing industrial process, the vegetal oils are heated at their melting temperature (approximately 40 °C) and mixed with the additives and emulsifiers to achieve a homogeneous mixture. In order to obtain a solid consistency of fat, the mixture is slowly cooled down and subjected to a hydrogenation process to produce fat saturation due to the unsaturated bonds of fatty acids of the vegetable oils breaking up [6].

The properties of margarines/spreads mainly depend on the characteristics of the vegetable oils, which are the major ingredients of the product, and the additives. The fat source is usually either soybean oil or sunflower oil blended with a hydrogenated vegetable oil, typically in the ratio 3:1 [7]. Other commodity vegetable oils include rape/canola, cottonseed, palm, palm kernel, and coconut, which may have been fractionated, blended, hydrogenated in varying degrees, and/or interesterified. Fish oil (hydrogenated or not) may also be included. Recent trends in the margarine market also consider mixtures with "healthy" vegetable oils with a low content of trans-saturated acid (e.g., high oleic sunflower or olive oils), as well as the addition of sterols. Regarding additives, these include surface-active agents, proteins, salt, and water, along with preservatives, flavours, and vitamins [8].

From the chemical point of view, the fat fraction of margarines/spreads is mainly composed of triacylglycerols (TAGs) [9]. Some data on the TAG composition of margarine/spread have been reported [10], but, as far as the authors are concerned, only one paper has been published, devoted to this matter [11]. The main compositional parameter being important for margarine/spread quality and healthy features is focused on the fatty acid (FA) profile [12–14]. The FA profile is commonly analysed using gas chromatography after derivatisation of them in the corresponding methyl esters. This approach provides incomplete chemical information and does not allow the knowledge of the variability linked to the combination of three FAs in the TAGs.

Conventional chromatographic techniques have been used successfully for the qualitative and quantitative determination of TAGs [15,16], mainly high-performance liquid chromatography (HPLC) in both normal and reverse phase modes [17–19]. The coupling of mass spectrometry to chromatographic instruments has drastically increased the analytical capabilities [20,21], and the regioisomeric and enantiomeric analysis of triacylglycerols is already affordable [22], but the chromatographic signals obtained from TAG profiling methods are never specific enough due to the great variety of isomers present in low proportion. In addition, 1H NMR has also been proven to be very useful in determining the composition in acyl groups of margarine samples, in only a few minutes and with minimal or no sample pretreatment [1,23]. Some examples of TAG profiling in some closely related margarine/spread products as milk and dairy foods can be found in references [24,25].

Nevertheless, the TAGs profile is characteristic of each vegetable oil, according to its botanical species, genus, or variety, which has a characteristic lipid profile. Consequently, TAGs profiling is both a useful and reliable tool in identifying vegetable oils and/or fraud detection [16] in order to verify the stated composition of margarines/spreads and authenticate them. Moreover, to our knowledge, there are no antecedents describing the comparison and classification of different margarines/spreads according to their geographical origin.

Another relatively alternative way to identify each vegetable oil using TAG analytical information is applying the fingerprinting methodology. This applies nonspecific instrumental signals where all the implicit, but nonevident, information contained in the analytical signal acquired from the samples is used, not being necessary to profile each chemical species present in the working solution. In this sense, signals coming directly from the measurement device or detector coupled with the chromatographic instrument are treated as a whole, and by means of advanced chemometric tools, as multivariate data analysis to extract, it is possible to reduce and process the extensive datasets in order to build proper multivariate models for classification or quantification purposes [26,27]. This methodology has become one of the most efficient and comprehensive methods to verify food identity [28].

This paper presents a multivariate qualitative analytical method for authenticating the geographical origin of margarines and fat-spread products. Several multivariate chemometrics tools were applied, such as principal components analysis (PCA), soft independent modelling by class analogy (SIMCA), and partial least squares-discriminant analysis (PLS-DA). As an analytical information source for building the multivariate models, both normal and reverse phase liquid chromatographic fingerprints were used. For this purpose, the analytical signals were acquired using a diode array detector (DAD) coupled with a high-performance liquid chromatographic (HPLC) system. Three strategies were tested to build the classification models: Two input-class, pseudo two input-class, and one input-class classification. The results from each classification method and strategy were compared and ranked on the basis of several classification performance metrics.

2. Materials and Methods

2.1. Chemicals and Samples

All solvents employed were HPLC-grade. Isopropanol and n-hexane were purchased from PANREAC Química (Barcelona, Spain), and acetonitrile was provided by VWR International Eurolab, S.L. (Barcelona, Spain).

A total of 35 margarine samples of different trade names or brands were analysed: 17 from Spain, 1 from France, 1 from Belgium, 1 from Germany, 1 from the Netherlands, 1 from the United Kingdom, and 13 from Morocco. Table 1 shows a description of the kind of vegetable oil employed in the manufacture of the products.

2.2. Sample Preparation

10% (*w/w*) solutions of margarine in n-hexane were prepared. The solutions were stirred for 5 min, then they were decanted and the supernatant was passed through a polytetrafluoroethylene (PTFE) membrane syringe filter (0.22 µm), and the resultant solutions were stored at −20 °C until analysis. Before the chromatographic analysis, the solutions were again diluted with n-hexane at a 1:1 ratio.

2.3. Instrumentation/Chromatography Conditions

Analysis using normal and reverse liquid chromatography coupled with a diode-array detector, (NP)HPLC-DAD and (RP)HPLC-DAD, respectively, was carried out with an Agilent 1260 series liquid chromatograph (Santa Clara, CA, USA), equipped with a column thermostat (Eppendorf CH30), a quaternary pump, and degasser auto sampler. Agilent ChemStation OpenLab CDS software (rev. C.01.09) for LC systems was used to collect and record data.

(NP)HPLC-DAD analysis was carried out using a column Lichrospher 100 CN (length 25 cm × i.d. 4 mm, particle size 4 µm) provided by Merck (Darmstadt, Germany). The column temperature was constant at 30 °C and the mobile phase was composed of n/hexane/isopropanol (96:4, *v/v*) at a flow rate of 1.2 mL min^{-1}. The run time was 29 min.

(RP)HPLC-DAD analysis was performed using the column DevelosilTM C30-UG-5 (length 25 cm × i.d. 4.6 mm, particle size 5 µm) from Nomura Chemical Co. (San Diego, CA, USA). During the analysis, the column temperature was at 50 °C. A mixture of acetronitre/isopropanol (40:60, *v/v*) was used as mobile phase at a flow rate of 1.2 mL min^{-1}. The chromatographic run time was 30 min.

The injection volume was 20 µL, and the DAD spectra were acquired at 210 and 254 nm. Figure 1a shows the fingerprint of a sample from Morocco recorded at 210 nm, and Figure 1b shows a sample from Spain that recorded the same wavelength obtained using (NP)HPLC-DAD. Likewise, Figure 1c,d displays the fingerprints of the same samples recorded at 210 nm using (RP)HPLC-DAD, respectively.

Table 1. Types of vegetable oils present in the samples.

Sample No.	Origin	Vegetable Oil
1		Sunflower, palm, and corn
2		Sunflower and palm
3		Sunflower and palm
4		Sunflower, coconut, and canola
5		Sunflower, linseed, coconut, and canola
6		Sunflower, linseed, and palm
7		Sunflower, linseed, palm, and canola
8		Olive, sunflower, linseed, and palm
9	Spain	Sunflower, linseed, and palm
10		Sunflower, linseed, and palm
11		Olive, sunflower, linseed, and palm
12		Soybean, sunflower, linseed, and palm
13		Sunflower and palm
14		Olive, sunflower, linseed, coconut, and shea
15		Sunflower, linseed, and palm
16		Sunflower, coconut, canola, and shea
17		Soybean, sunflower, linseed, and palm
18		-
19		-
20		Soybean
21		-
22		-
23		Soybean and corn
24	Morocco	Soybean
25		-
26		Corn
27		Soybean
28		Soybean
29		Soybean
30		Soybean
31	The Netherlands	Sunflower, palm, and coconut
32	United Kingdom	Sunflower, palm, and rapeseed
33	France	Rapeseed, palm, olive, and sunflower
34	Germany	Soybean, palm, rapeseed, and coconut
35	Belgium	Palm, coconut, and canola

The hyphen "-" signifies that the kind of vegetable oil present in the samples is unknown.

With respect to the TAG composition of the blended fat, the obtained fingerprint depends on the proportion of unsaturated/saturated FAs. Note that only the unsaturated FAs generate a measurable signal as the saturated FAs are almost transparent to the UV-absorption detector at the working concentration. Consequently, fingerprints show specificity from the distribution of the unsaturated FAs into the different TAGs. This characteristic causes the UV-absorption fingerprints to be different to the fingerprint acquired from other more universal detectors used in edible fat analyses, such as refractive index (RID), evaporative light scattering (ELSD), and corona charged aerosol (CAD).

2.4. Chemometrics

The raw data files from each chromatogram were exported in "comma separated value" (CSV) format and then turned to "xls" format (Microsoft Excel). For (NP)HPLC-DAD and (RP)HPLC-DAD analysis, a data vector composed of 4500 and 4350 variables (each one a specific absorbance to a scanned wavelength) was collected for each sample, respectively. The chromatographic fingerprints were reproducible from sample to sample, and thus, no alignment process was applied. The only preprocessing data carried out was the mean centring of the dataset before the development of the models. PCA, SIMCA, and partial least squares-discriminant analysis (PLS-DA) methods were built using The Unscrambler software ver. 9.7 (CAMO, Oslo, Norway).

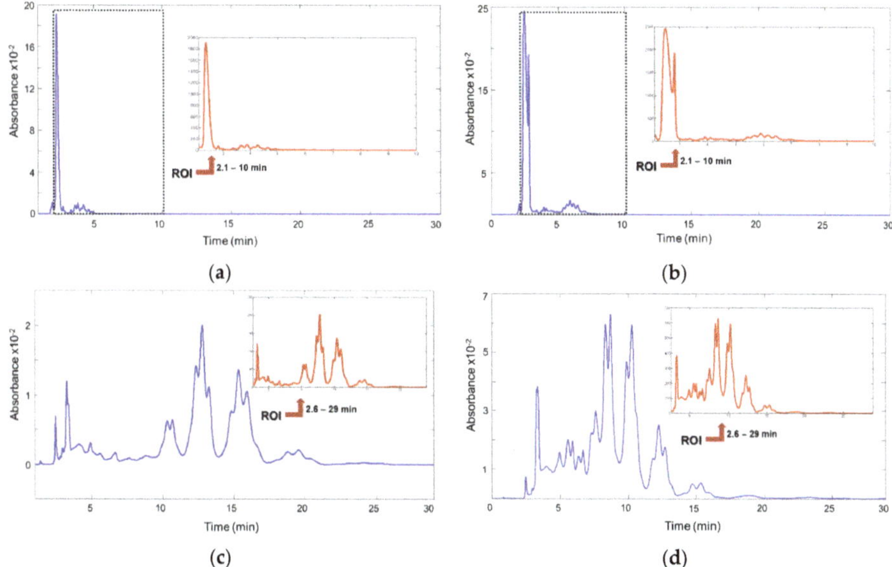

Figure 1. Chromatogram of a margarine/spread samples showing the region of interest used to build the classification models: (**a**,**b**) Normal phase from Morocco and Spain, respectively; (**c**,**d**) reverse phase from Morocco and Spain, respectively.

Three classification strategies were applied: one input-class (1iC), two input-class (2iC), and pseudo two input-class (*p*2iC) to develop the different Spain/Morocco binary classification methods. Two input-class is the conventional binary classification methodology in which the model is trained using samples from target and nontarget class. One input-class strategy is used only in class modelling methods such as SIMCA. In these, the model is trained only with the target class. This methodology presents an advantage in food authentication as it is only necessary to analyse samples from target class (genuine class). Nevertheless, the modelling methods are less reliable than discriminant analysis methods; thus, the pseudo two input-class is applied in order to employ discriminant methods when it is only possible to have samples from the "genuine class". A more detailed description about these strategies is shown in the paper published by Jiménez-Carvelo et al. [29]. For each strategy, the original data set was randomly split into different sets: Training and validation set. For 2iC, the training set was made up of 20 margarine samples, and the external validation set was composed of the remaining margarine samples. For *p*2iC, the training set was made up of 10 samples and 6 solvent analytical blank replicates, and the validation set was composed of 20 samples. For 1iC, the training set was composed of 10 samples, and the validation set included 20 samples. Table 2 details the sample distribution regarding the geographical origin for each classification strategy.

Once all the classification models were validated, they were critiqued using the samples from the validation set, and new classification models were built. To follow, these final models were used to predict the continent origin of the margarine/spread samples from the European countries other than Spain (France, Belgium, Germany, the Netherlands, and the United Kingdom). These samples constitute the prediction set.

Table 2. Distribution of the samples used in the different classification datasets regarding geographical origin.

		SIMCA		PLS-DA	
Dataset	Origin	(1ic)	(2ic)	(*p*2ic)	(2ic)
	Spain	10	10	10	10
Training set	Morocco	–	10	–	10
	Blank	–	–	6	–
	Spain	7	7	7	7
Validation set	Morocco	13	3	13	3
Prediction set	Europe (other than Spain)	–	–	–	5

3. Results and Discussion

Firstly, four PCA models were built using the dataset composed of the whole fingerprint from each sample in both normal and reverse phase modes. Table 3 shows the number of PCs chosen for each model. Figure 2 shows the scores on the PC1–PC2 plane of the fingerprints acquired at 210 nm and 254 nm. The best groupings were found in the PCA models from data collected at 210 nm.

Table 3. Characteristics of the principal components analysis (PCA) models.

	(RP)HPLC-DAD		(NP)HPLC-DAD	
Wavelength	PCs	% Var	PCs	% Var
210 nm	4	90.00	4	95.00
254 nm	4	99.00	4	93.00

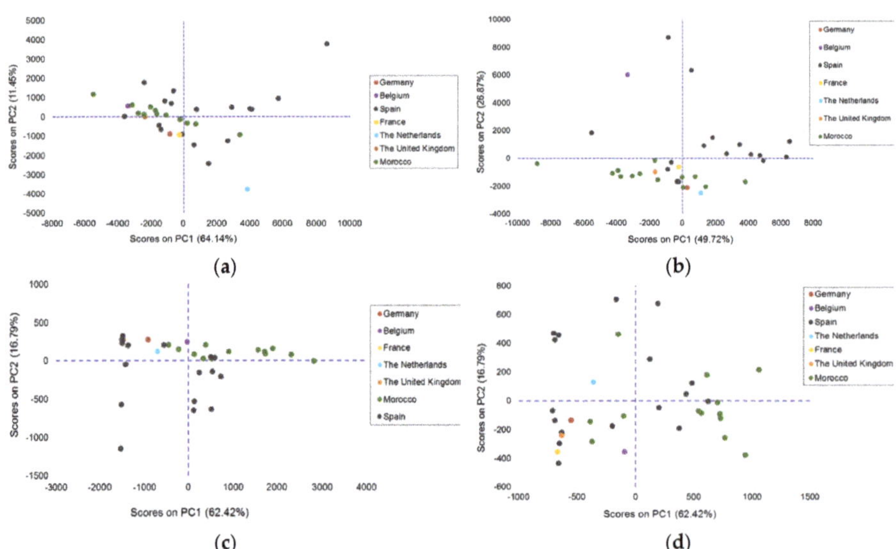

Figure 2. PCA scores obtained from the fingerprint data of the 35 margarine samples: PC1–PC2 plane of the chromatogram acquired at 210 nm: (**a**) In reverse phase; (**b**) In normal phase; and acquired at 254 nm: (**c**) In reverse phase; (**d**) In normal phase.

Once PCA models were evaluated, the three strategies (2iC, 1iC, and *p*2iC) were applied to develop the SIMCA and PLS-DA classification models. Nevertheless, the best results were found using the 2iC strategy from the 210-collecting dataset for both normal and reverse phases. Thus, the results shown in the following sections correspond with these datasets, considering the "Spanish class" as the target class in every case. It is important to emphasise that only the values of the performance metrics related to the target class provide useful information when the classification is used as a screening method, because the errors of the samples belonging to the nontarget class are not critical information as they all may be subjected to the confirmatory method.

3.1. SIMCA Methods

The application of SIMCA involves building a classification method in which each class of the training set is modelled independently (Spain model and Morocco model). Once the individual PC models were built, these were assembled to perform the classification of the samples. Two SIMCA models, for both the normal phase and reverse phase dataset, were then developed choosing four principal components (PCs).

The classification results were evaluated, attending to Coomans' plot. Figure 3 displays the Coomans' plot of the two SIMCA models. It can be observed that some samples are located in the bottom-left quadrant; thus, these samples were considered inconclusive and they were not taken into account for estimation of the quality metrics for each model. Table 4 shows the results of the success/error contingencies, and Table 5 collects the quality metrics calculated for each model.

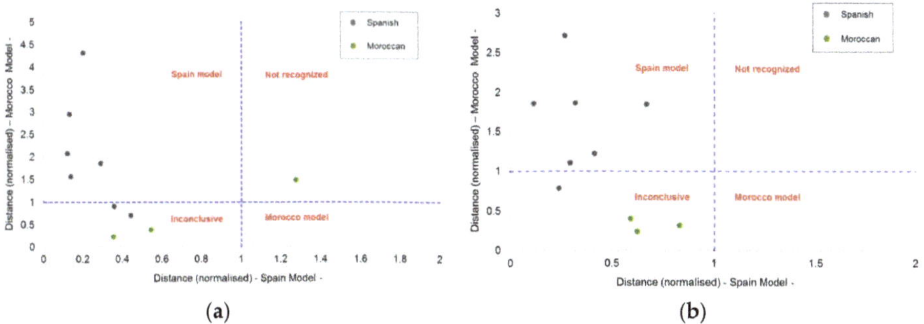

Figure 3. Coomans' plot from (**a**) normal and (**b**) reverse liquid chromatography coupled with diode-array detector ((NP)HPLC-DAD and (RP)HPLC-DAD, respectively) datasets at 210 nm.

Table 4. Soft independent modelling of class analogy (SIMCA) classification results for the validation set.

(NP)HPLC-DAD

Assignation		7	3	10
	O	0	1	1
	I	2	2	4
	nT	0	0	0
	T	5	0	5
		T	nT	

Actual

(RP)HPLC-DAD

Assignation		7	3	10
	O	0	0	0
	I	1	3	4
	nT	0	0	0
	T	6	0	6
		T	nT	

Actual

T: Target class (Spanish class); nT: Nontarget class (Moroccan class); I: Inconclusive samples; O: Samples not considered as belonging to any class.

Table 5. Quality metrics of the SIMCA models.

Parameter	(NP)HPLC-DAD	(RP)HPLC-DAD
Sensitivity (or Recall)	0.71	0.86
Specificity	0.00	0.00
Positive predictive value (Precision)	1.00	1.00
Negative predictive value	–	–
Youden index	−0.29	−0.14
Positive likelihood rate	0.71	0.86
Negative likelihood rate	–	–
F-measure	0.83	0.92
Discriminant power	–	–
Efficiency (or Accuracy)	0.50	0.60
AUC (Correctly classified rate)	0.36	0.43
Matthews correlation coefficient	–	–
Kappa coefficient	0.23	0.31

The hyphen "-" signifies that the performance feature cannot be determined.

Coomans' plot is a tool to graphically visualize principal groupings results in pairwise plots easily, in which the two axes represent the normalised orthogonal distances of all the samples with respect to each individual model. Ideally, the validation samples should be classified in one class or another (target or nontarget class). In real conditions, some validation samples could be assigned to both classes simultaneously, as these samples are considered inconclusive ones, or to neither of them (the samples are not recognized as belonging to any class).

3.2. PLS-DA Methods

The discriminant methods are generated through establishing the boundaries for the different categories defined by the training objects. PLS-DA is a latent variable-based method whose development involves two stages: (I) Firstly, a PLS regression model is established from the latent variables (LV) to establish limits between the classes, and then, (ii) a discriminant analysis (DA) is performed to classify the samples into a specific class.

Two PLS-DA models were built using five LVs from both normal and reverse liquid chromatography datasets. The "Spanish class" was defined by values equal to 0, while the "Moroccan class" was defined by a value of 1. The decision criterion established for the classification of the samples was a threshold value of 0.5, i.e., all the margarine samples with scores greater than 0.5 were classified to the Moroccan class, and margarine samples with scores lower than 0.5 were assigned to the Spanish class. In addition, for the purpose of improving the reliability of the validation and prediction results, an uncertainty interval was established as plus/minus 0.1 the settled threshold value for the training samples.

Figure 4 shows the classification plots obtained from PLS-DA methods. The blue line displays the classification threshold, and the orange strip represents the uncertainty region, within which any sample is stated as inconclusive.

The PLS-DA classification performances were evaluated, calculating the same quality metrics as SIMCA. These were estimated using the success/error contingency for each class in which samples of the validation set were arranged. Both contingency tables and quality metrics for the two PLS-DA classifiers established are shown in Tables 6 and 7, respectively.

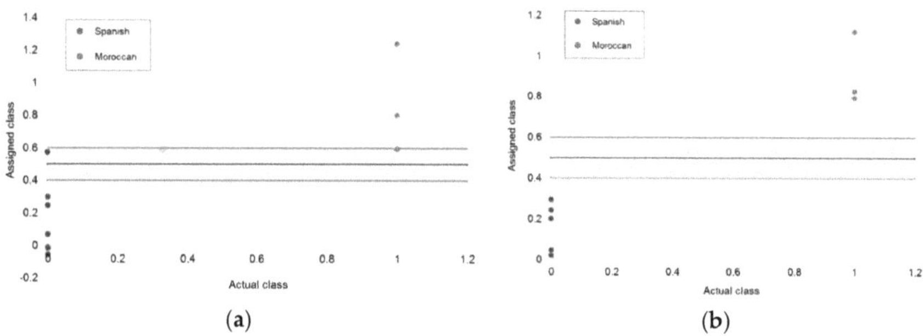

(a)   (b)

Figure 4. Classification plot of the partial least squares-discriminant analysis (PLS-DA) models from liquid chromatography (LC) data obtained in (**a**) normal phase and (**b**) reverse phase at 210 nm. The orange area represents the inconclusive area.

Table 6. PLS-DA classification results for the validation set.

(NP)HPLC-DAD

		7	**3**	**10**
	O	0	0	0
	I	1	1	2
Assignation	**nT**	0	2	2
	T	6	0	6
		T	**nT**	

Actual

(RP)HPLC-DAD

		7	**3**	**10**
	O	0	0	0
	I	0	0	0
Assignation	**nT**	0	3	3
	T	7	0	7
		T	**nT**	

Actual

T: Target class (Spanish class); nT: Nontarget class (Moroccan class); I: Inconclusive samples; O: Samples not considered as belonging to any class.

Table 7. Quality metrics of the PLS-DA models.

Parameter	(NP)HPLC-DAD	(RP)HPLC-DAD
Sensitivity (or Recall)	0.86	1.00
Specificity	0.67	1.00
Positive predictive value (Precision)	1.00	1.00
Negative predictive value	1.00	1.00
Youden index	0.52	1.00
Positive likelihood rate	2.57	–
Negative likelihood rate	0.21	0.00
F-measure	0.92	1.00
Discriminant power	0.60	–
Efficiency (or Accuracy)	0.80	1.00
AUC (Correctly classified rate)	0.76	1.00
Matthews correlation coefficient	0.76	1.00
Kappa coefficient	0.62	1.00

The hyphen "-" is signifying that the performance feature cannot be determined.

As can be seen, PLS-DA models provided better classification results than SIMCA ones. In particular, the model developed using the (RP)HPLC-DAD dataset was the best, as all the quality metrics were equal to 1.00 and there were not any samples classified as inconclusive. All the margarine/spread samples from the target class were well classified (probability = 0), and the samples from the nontarget class were also classified correctly (probability = 1).

As stated in Section 2.4, once SIMCA and PLS-DA models were correctly validated, the best model was applied to predict the class similarity of the margarine/spread samples from Belgium, France, Germany, the Netherlands, and the United Kingdom. For this purpose, the PLS-DA model built from the (RP)HPLC-DAD dataset was employed. The main goal of performing this classification was to test: (i) If the model was able to classify these samples as inconclusive as they belonged to neither Spanish nor Moroccan classes, and (ii) if the model was able to find similarities with any of the modelling classes (originating from Spain or Morocco). As can be seen in Figure 5, the model classified the five samples in the Spanish class. Probably, the overriding reason is related to the manufacturing process, because in Europe, there is an extensive and descriptive legislation that the margarine/spread products should be similar enough and all alternative to those manufactured in Morocco.

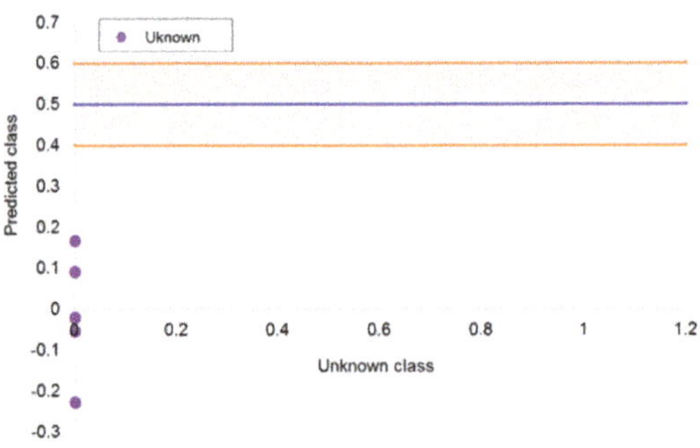

Figure 5. Class predictions plot for the samples from the European countries other than Spain.

4. Conclusions

A methodology for the discrimination of margarine and fat-spread foodstuffs from Spain and Morocco using the fingerprinting methodology was described. Normal and reverse phase liquid chromatography coupled with UV-absorption was used as an analytical technique to acquire the 210 and 254 nm chromatographic fingerprints. In addition, different multivariate classification methods and strategies were tested, and the 210 nm PLS-DA models were found to perform better than the SIMCA ones, as all the samples were correctly classified.

It has to be stressed that the proposed methodology could also be applied to differentiate margarine and fat-spread products produced in any European country from the ones manufactured in Morocco, because all the Europe samples gave similar results of belonging to the Spanish class. Even so, further studies are necessary to test this hypothesis, which is currently being performed by the authors.

Author Contributions: Conceptualization, M.N. and B.S.; Formal analysis, A.M.J.-C., and L.C.-R.; Investigation, M.G.B.-G.; Methodology, S.B.; Project administration, M.N., M.G.B.-G., and B.S.; Resources, M.G.B.-G. and L.C.-R.; Software, S.B.; Supervision, L.C.-R.; Validation, S.B.; Writing—original draft, A.M.J.-C.; Writing—review and editing, A.M.J.-C. and L.C.-R.

Funding: This research received no external funding.

Acknowledgments: The authors are very grateful to all the study participants.

Conflicts of Interest: The authors declare no conflict of interest.

References

1. Guillén, M.D.; Ibargoitia, M.L.; Sopelana, P. Margarine: Composition and analysis. In *Encyclopedia of Food and Health*; Caballero, B., Finglas, P.M., Toldrá, F., Eds.; Academic Press: Oxford, UK, 2016; Volume 3, pp. 646–653. [CrossRef]
2. *Codex Stan 256-2007 Standard for Fat Spreads and Blended Spreads*; Codex Alimentarius FAO/WHO: Rome, Italy, 2017. [CrossRef]
3. Gunstone, F.D. *Oils and Fats in the Food Industry*; Blackwell Publishing: Oxford, UK, 2008; Chapter 8; pp. 113–136.
4. *Regulation (EU) No 1308/2013 Establishing a Common Organization of the Markets in Agricultural Products*; OJ 02013R1308-EN-003.001-223 (Consolidated Version 01.08.2017); EU: Brussels, Belgium, 2017.
5. *Décret n° 1153-55 du 18 kaada 1389 (26 Janvier 1970) Portant Réglementation pour L'application du Dahir du 23 Kaada 1332 (14 Octobre 1914) sur la Répression des Fraudes en ce qui Concerne la Fabrication et la Vente de la Margarine, B.O n° 2988 du 04/02/1970*; Government Printing Office: Fait à Rabat, Morocco, 1970; p. 214. (In French)
6. Wassell, P. Bakery fats. In *Fats in Food Technology*, 2nd ed.; Rajah, K.K., Ed.; John Wiley & Sons: Oxford, UK, 2014; pp. 40–81.
7. Keogh, M.K. Chemistry and technology of butter and milk fat spreads. In *Advanced Dairy Chemistry: Lipids*, 3rd ed.; Fox, P.F., McSweeney, P.L.H., Eds.; Springer: New York, NY, USA, 2006; Volume 2, pp. 333–363.
8. Gunstone, F.D. *The Chemistry of Oils and Fats: Sources, Composition, Properties and Uses*; Blackwell Publishing: Oxford, UK, 2004; Chapter 10; pp. 238–256.
9. Marini, F. Triacylglycerols: Characterization and determination. In *Encyclopedia of Food and Health*; Caballero, B., Finglas, P.M., Toldrá, F., Eds.; Academic Press: Oxford, UK, 2016; Volume 5, pp. 345–350. [CrossRef]
10. List, G.R.; Steidley, K.R.; Neff, W.E. Commercial spreads formulation, structures and properties. *Inform* **2000**, *11*, 980–986.
11. Craig Byrdwell, W.; Neff, W.E.; List, G.R. Triacylglycerol analysis of potential margarine base stocks by high-performance liquid chromatography with atmospheric pressure chemical ionization mass spectrometry and flame ionization detection. *J. Agric. Food Chem.* **2001**, *49*, 446–457. [CrossRef] [PubMed]
12. Larqué, E.; Garaulet, M.; Pérez Llamas, F.; Zamora, S.; Tebar, F.J. Fatty acid composition and nutritional relevance of most widely consumed margarines in Spain. *Grasas y Aceites* **2003**, *54*, 65–70. [CrossRef]
13. Kroustallaki, P.; Tsimpinos, G.; Vardavas, C.I.; Kafatos, A. Fatty acid composition of Greek Margarines and their change in fatty acid content over the past decades. *Int. J. Food Sci. Nutr.* **2011**, *62*, 685–691. [CrossRef] [PubMed]
14. Garsetti, M.; Balentineb, D.A.; Zock, P.L.; Blom, W.A.M.; Wanders, A.J. Fat composition of vegetable oil spreads and margarines in the USA in 2013: A national marketplace analysis. *Int. J. Food Sci. Nutr.* **2016**, *67*, 372–382. [CrossRef] [PubMed]
15. Buchgraber, M.; Ulberth, F.; Emons, H.; Anklam, E. Triacylglycerol profiling by using chromatographic techniques. *Eur. J. Lipid Sci. Technol.* **2004**, *106*, 621–648. [CrossRef]
16. Indelicato, S.; Bongiornob, D.; Pitonzo, R.; Di Stefano, V.; Calabrese, V.; Indelicato, S.; Avellone, G. Triacylglycerols in edible oils: Determination, characterization, quantitation, chemometric approach and evaluation of adulterations. *J. Chromatogr. A* **2017**, *1515*, 1–16. [CrossRef] [PubMed]
17. Lisa, M.; Lynen, F.; Holcapek, M.; Sandra, P. Quantitation of triacylglycerols from plant oils using charged aerosol detection with gradient compensation. *J. Chromatogr. A* **2007**, *1176*, 135–142. [CrossRef] [PubMed]
18. Rombaut, R.; De Clercq, N.; Foubert, I.; Dewettinck, K. Triacylglycerol analysis of fats and oils by evaporative light scattering detection. *J. Am. Oil Chem. Soc.* **2009**, *86*, 19–25. [CrossRef]
19. de la Mata Espinosa, P.; Bosque Sendra, J.M.; Cuadros Rodríguez, L. Quantification of triacylglycerols in olive oils using HPLC-CAD. *Food Anal. Methods* **2011**, *4*, 574–581. [CrossRef]
20. Laakso, P. Mass spectrometry of triacylglycerols. *Eur. J. Lipid Sci. Technol.* **2002**, *104*, 43–49. [CrossRef]
21. Xu, S.-L.; Wei, F.; Xie, Y.; Xin, L.; Dong, X.; Chen, H. Research advances based on mass spectrometry for profiling of triacylglycerols in oils and fats and their applications. *Electrophoresis* **2018**, *39*, 1558–1568. [CrossRef] [PubMed]

22. Rezanka, T.; Pádrová, K.; Sigler, K. Regioisomeric and enantiomeric analysis of triacylglycerols. *Anal. Biochem.* **2017**, *524*, 3–12. [CrossRef] [PubMed]

23. Sopelana, P.; Arizabaleta, I.; Ibargoitia, M.L.; Guillén, M.D. Characterisation of the lipidic components of margarines by 1H nuclear magnetic resonance. *Food Chem.* **2013**, *141*, 3357–3364. [CrossRef] [PubMed]

24. Fontecha, J.; Juárez, M.; de la Fuente, M.A. Triacylglycerols in dairy foods. In *Handbook of Dairy Foods Analysis*; Nollet, L.M.L., Toldra, F., Eds.; CRC Press: Boca Raton, FL, USA, 2010; pp. 169–187.

25. Omar, K.A.; Gounga, M.E.; Liu, R.; Mwinyi, W.; Aboshora, W.; Ramadhan, A.H.; Sheha, K.A.; Wang, X. Triacylglycerol composition, melting and crystallization profiles of lipase catalysed anhydrous milk fats hydrolysed. *Int. J. Food Prop.* **2017**, *20*, S1230–S1245. [CrossRef]

26. Bosque Sendra, J.M.; Cuadros Rodríguez, L.; Ruiz Samblás, C.; de la Mata, A.P. Combining chromatography and chemometrics for the characterization and authentication of fats and oils from triacylglycerol compositional data—A review. *Anal. Chim. Acta* **2012**, *724*, 1–11. [CrossRef] [PubMed]

27. Cuadros Rodríguez, L.; Ruiz Samblás, C.; Valverde Som, L.; Pérez Castaño, E.; González Casado, A. Chromatographic fingerprinting: An innovative approach for food 'identitation' and food authentication—A tutorial. *Anal. Chim. Acta* **2016**, *909*, 9–23. [CrossRef] [PubMed]

28. Bagur-González, M.G.; Pérez-Castaño, E.; Sánchez-Viñas, M.; Gázquez-Evangelista, D. Using the liquid-chromatographic-fingerprint of sterols fraction to discriminate virgin olive from other edible oils. *J. Chromatogr.* **2015**, *1380*, 64–70. [CrossRef] [PubMed]

29. Jiménez Carvelo, A.M.; Pérez Castaño, E.; González Casado, A.; Cuadros Rodríguez, L. One input-class and two input-class classifications for differentiating olive oil from other edible vegetable oils by use of the normal-phase liquid chromatography fingerprint of the methyl-transesterified fraction. *Food Chem.* **2017**, *221*, 1784–1791. [CrossRef] [PubMed]

Article

Differentiation between Ripening Stages of Iberian Dry-Cured Ham According to the Free Amino Acids Content

Ángela Alcazar Rueda [1], José Marcos Jurado [1], Fernando de Pablos [1] and Manuel León-Camacho [2,*]

[1] Department of Analytical Chemistry, Faculty of Chemistry, University of Seville, calle Profesor García González 1, E-41012 Seville, Spain; alcazar@us.es (A.A.R.); jmjurado@us.es (J.M.J.); fpablos@us.es (F.d.P.)
[2] Lipid Characterization and Quality Department. Instituto de la Grasa (C.S.I.C.), Ctra. Utrera km 1, Campus Universitario Pablo de Olavide, Edificio 46, E-41013 Seville, Spain
* Correspondence: mleon@ig.csic.es; Tel.: +34-954611550

Received: 19 December 2019; Accepted: 9 January 2020; Published: 12 January 2020

Abstract: In this paper, the differentiation of three ripening stages, postsalting, drying, and cellar, of Iberian dry-cured ham has been carried out according to their free amino acids contents. Eighteen L-amino acids, alanine, 2-aminobutanoic acid, aspartic acid, cysteine, glutamine, glycine, histidine, hydroxyproline, isoleucine, leucine, lysine, methionine, phenylalanine, proline, serine, threonine, tyrosine, and valine have been determined by gas chromatography with derivatization with N,O-*bis*(trimethylsilyl)-trifluoroacetamide. Gas chromatography–mass spectrometry was used to confirm the presence of the eighteen amino acids in the ham samples, and gas chromatography using a DB-17HT column and flame ionization detector was used for quantitative determination. Extraction with a mixture methanol-acetonitrile has been carried out, achieving recoveries in the range 52–164%. Methimazole was used as internal standard. Limits of detection ranged between 7.0 and 611.7 mg·kg^{-1}. Free amino acids have been used as chemical descriptors to differentiate between the ripening stages. Principal component analysis and linear discriminant analysis have been used as chemometric techniques, achieving complete differentiation between the ripening stages. Alanine, tyrosine, glutamine, proline, 2-aminobutanoic acid, cysteine, and valine were the most differentiating amino acids.

Keywords: free amino acids; gas chromatography; mass spectrometry; Iberian dry-cured ham; food authentication; chemometrics; pattern recognition

---

1. Introduction

Iberian dry-cured ham is a meat product manufactured following traditional methods [1]. It is highly appreciated by the consumers, and there is an increasing demand in the last years because of its good culinary qualities. Additionally, animal husbandry is done outdoors, improving the animal welfare, reducing the environmental impact, and protecting the ecosystem [2,3]. Iberian dry-cured ham has a preponderant role in the economies of several Spanish production areas, and various Protected Designations of Origin (P.D.O.) at the southwest of the country have been established, being P.D.O. "Los Pedroches" one of them [4]. Traditionally, the production of dry-cured ham consists of salting, postsalting, drying, and a final stage in a cellar. During the postsalting, drying, and cellar stages, ripening of the ham takes place. The time and duration of the postsalting, drying, and cellar stages vary depending on the type of dry-cured ham. In addition to lipolysis and oxidation, during ripening, endogenous enzymes degrade lipids and proteins to fatty and amino acids, which have an important influence on the flavor of dry-cured ham [5,6]. The time for the drying and cellar stages has an important influence on the quality of the final product. It takes from 23 months to 2–3 years for the highest-quality dry-cured hams [7]. A longer time of ripening leads to greater enzymatic degradation with an important effect on taste and flavor, producing a higher-quality dry-cured ham [8]. For this reason, the characterization of dry-cured ham and studies on their chemical evolution related to taste and flavor qualities are of great interest. In this way, various fractions have been studied, such as the volatile and the lipidic fractions of subcutaneous fat [9–12].

In the decade of the 1950s, amino acids have been determined by ion-exchange chromatography followed by post-column derivatization with ninhydrin and UV detection [13,14]. Reversed-phase high-performance liquid chromatography (HPLC) has been used for amino acid determination in the last thirty years, due to the flexibility of the technique and giving the possibility of using fluorescence detection and a higher sensitivity [15]. Analytical methods based on mass spectrometry (MS) and tandem MS (MS/MS) have also been reported for the determination of amino acids in foods [16,17]. In recent years, gas chromatography (GC) combined with flame ionization detector (FID) and/or mass spectrometer (MS) has been more commonly applied to the analysis of amino acids [18–24]. GC methods require a derivatization step to increase the volatility of the amino acids, being the methods based on the formation of silyl derivatives more commonly used [25,26].

Till now, the amino acids have not been considered for the differentiation of the ripening stage in the elaboration process of Iberian dry-cured ham. In this paper, the free L-amino acids alanine, 2-aminobutanoic acid, aspartic acid, cysteine, glutamine, glycine, histidine, hydroxyproline, isoleucine, leucine, lysine, methionine, phenylalanine, proline, serine, threonine, tyrosine, and valine have been determined by gas chromatography in Iberian dry-cured ham at postsalting, drying, and cellar ripening stages. Those amino acids have been used as chemical descriptors to differentiate between the three ripening stages. Principal component analysis (PCA) and linear discriminant analysis (LDA) have been used as pattern recognition methods.

2. Materials and Methods

2.1. Reagents and Standards

All solvents employed were HPLC-grade. N-Hexane and multisolvent TM HPLC ACS grade were purchased from VWR (Barcelona, Spain), and diethyl-ether, methanol, acetonitrile, and anhydrous pyridine were provided by Merck (Darmstadt, Germany). L-Amino acids standards: alanine, sarcosine, glycine, aminobutiric acid, β-alanine, valine, leucine, proline, isoleucine, serine, threonine, methionine, aspartic acid, hydroxyproline, cysteine, ornithine, citruline, arginine, phenyl alanine, glutamic acid, asparagine, homoarginine, lysine, histidine, glutamine, tyrosine, tryptophan, cistine, and hydroxytryptophan were supplied by Sigma Chemical Co. (St. Louis, MO, USA). Methimazole was used as internal standard and it was obtained from Sigma Chemical Co. (St. Louis, MO, USA). N,O-*bis*(trimethylsilyl)-trifluoroacetamide (BSTFA) from Sigma Chemical Co. (St. Louis, MO, USA) was used as derivatizing reagent. Other reagents were of analytical grade.

2.2. Samples and Sample Treatment

Hams from the protected designation of origin "Los Pedroches" were obtained from five Iberian 18-month-old pigs, fattened extensively with acorns, and pastured for 110 days prior to slaughter and processed in an industry for 26 months. In the following, the stages and the number of days since the beginning of the process are described. After the slaughter, hams were removed from the carcasses after 24 h and storage at 1 °C. Then they were placed in piles completely covered with marine salt with no contact between each other during one day for each kilogram of the ham. Next, the hams were hung at postsalting period during 90 days under controlled temperature and humidity, and then they were taken to a dryer at room temperature for 230 days. After this period, hams were left to mature during 400 days in a cellar. Consequently, the three steps considered in the process were postsalting, drying, and cellar. To cover the three steps of the process, samples were taken at 0, 60, 115, 180, 325,

440, 556, 653, and 783 days from the beginning of the postsalting step. The samples were taken from muscle, *biceps femoralis* and *cuadriceps femoralis,* and kept at −32 °C until analysis.

2.3. Instrumentation and Gas Chromatography Analysis

A Varian 3800 gas chromatograph equipped with a fused silica capillary column of 30 m × 0.25 mm internal diameter coated with a 0.15 μm film thickness of DB-17HT (J&W Scientific, Albany, NY, USA) stationary phase, a flame ionization detector (FID), and a Varian 8400 automatic injector was used. The oven temperature was kept at 85 °C for 2 min, and it was then raised to 100 °C at a rate of 1 °C min^{-1}. Next, it was then raised to 258 °C at a rate of 6 °C min^{-1}. The injector temperature was kept at 290 °C, while the detector temperature was fixed at 310 °C. Hydrogen at a constant flow of 1.0 mL min^{-1} was used as carrier gas with a split ratio of 1:10. Air and hydrogen with flow rates of 300 and 30 mL min^{-1}, respectively, were used for the detector, which had an auxiliary flow of 30 mL min^{-1} of nitrogen.

GC–MS was applied to identify amino acids in the samples. A Varian CP3800 gas chromatograph coupled to a Saturn 2000 ion trap mass spectrometer (Varian, Palo Alto, CA, USA) equipped with a CP8400 autosampler was used. A DB-5MS (J&W Scientific, Albany, NY, USA) fused silica capillary column of 30 m × 0.25 mm i.d., coated with a 0.25 μm film thickness of DB-5MS stationary phase, was used. The oven temperature was initially kept at 90 °C for 2 min, and it was then raised to 120 °C at a rate of 2 °C min^{-1}. Next, it was then raised to 258 °C at a rate of 5 °C min^{-1}, followed by an isothermal period of 5 min at the latter temperature. The injector temperature was 290 °C. Hydrogen was used as carrier gas at 1.0 mL min^{-1} in constant flow mode and a split ratio of 1:20. The MS detector was operated in full scan mode from 25 to 650 amu at 1 scan/sec. Ion source and transfer line temperatures were kept 200 and 290 °C, respectively. The electron energy was 70 eV, a resolution of 1 and the emission current 10 μA were fixed. Dwell time and inter-channel delay were 0.08 s and 0.02 s, respectively. Varian Mass Spectrometry Workstation version 6.30 software (Varian, Palo Alto, CA, USA) was used for data acquisition and processing of the results.

The free amino acids content in the samples was determined according to the following procedure. Muscle samples of 0.5 g were cut up into small pieces and homogenized. Following, the samples were degreased with 3 × 10 mL of n-hexane-diethyl-ether (4:1 *v/v*) solvent extraction using a vortex agitator. Then, 1 mL of methanol containing 2.4 mg mL^{-1} of methimazole as internal standard was added and subsequently dried in a rotary evaporator at 30 °C under reduced pressure. The obtained residue was extracted with 3 × 5 mL of methanol-acetonitrile (1:1 *v/v*) using a vortex agitator. Of the solution obtained, 1 mL was filtered and next, evaporated to dryness in a rotary evaporator at 30 °C under reduced pressure. Acetonitrile (0.3 mL) and 0.3 mL of the derivatizing reagent (BSTFA) were added, and the mixture was heated at 80 °C during 30 min to obtain the trimethylsilyl derivatives. Of this solution, 1 μL was injected into the gas chromatograph. Triplicate analyses were performed.

2.4. Chemometrics

For chemometric calculations, a data matrix was prepared. The analyzed amino acids are used to describe the ham samples. Pattern recognition methods were applied to the data matrix, composed of 18 columns that correspond to the analyzed amino acids and 45 rows that are the ham samples. The data were analyzed using Statistica 8.0 software (Statsoft Inc., Tulsa, OK, USA).

Chemometrics is applied with several purposes. By one side, we try to visualize tendencies of the samples along the ripening process. On the other side, we are looking for appropriate classification rules to differentiate between postsalting, drying, and cellar stages. Additionally, information about the discriminant capacity of the variables can be obtained.

Tendencies of the samples can be studied by using principal component analysis (PCA). PCA obtains new variables as linear combinations of the variables that, in this case, are the determined amino acids. The new variables are called principal components (PCs). There is no correlation between these PCs [27,28]. They are obtained in a sequential way, and each successive PC considers the remaining variability. The information provided by the determined variables is condensed in first PCs. Usually, the two first PCs, PC1 and PC2, account for an important part of the information, and plots using PC1 and PC2 as variables (scores plot) are very useful to visualize the trends of the data matrix. [29].

A classification rule was obtained by applying linear discriminant analysis (LDA). Calculations carried out with LDA produce discriminant functions (DFs) that are obtained as linear combinations of the variables which best separate the three considered ripening stages of the process. LDA is a hard modelling technique because the memberships of every sample and the number of classes or groups of samples have to be previously known. In this case, three classes, postsalting, drying, and cellar classes, are considered and two DFs have been calculated.

3. Results and Discussion

3.1. Determination of Free Amino Acids by Gas Chromatography

Chromatographic analysis provides rapid and reliable separation of chemically similar compounds in complex food matrices. Dry-cured ham includes compounds with a wide range of polarities, some of them have low polarity, like lipids, and others are strongly polar, like amino acids [30]. GC is a high-resolution technique very useful to afford the analysis of this type of sample. Due to the low volatility, derivatization of free amino acids is performed for their analysis by GC. Several methods have been proposed in the literature. Some of them are based on multi-step procedures, which involve esterification of the carboxyl group followed by acylation of the remaining functional groups, and others are based on the formation of the silyl ethers [31]. Several derivatizing reagents were tested, including BSTFA, N,O-bis(trimethyldilyl)acetamide (TRI-SIL/BSA), and trimethyl-chlorosilane/hexamethyldisilazane/pyridine. The most effective derivatization was achieved when using BSTFA, while TRI-SIL/BSA and trimethyl-chlorosilane/hexamethyldisilazane/pyridine showed a very low efficiency. The use of a bulkier silylating group avoided the inconvenience of multiple derivative formation observed with some amino acids [31]. In order to improve the reaction efficiency and due to the acid hydrolysis reaction that takes place, small amounts of acetonitrile were added. Optimization of temperature and time of reaction was carried out, and the final optimal conditions were as follows: addition of 0.3 mL of acetonitrile and 0.3 mL of the derivatizing reagent (BSTFA) at 80 °C during 30 min.

Several types of columns have been used to perform the analysis of amino acids by GC, some of them being low-polarity columns of methylpolysiloxane or silicone phases, methyl 5% phenylpolisiloxane, and others medium-polarity columns with methyl 50% phenyl polysiloxane [26,32]. In this work, a high temperature column with methyl 50% phenyl polysiloxane stationary phase DB-17HT has been used for the determination of the amino acids. Figure 1 shows a chromatogram of standards obtained using FID detector. The relative retention times of the amino acids to L-methionine are included in Table 1. In general, a good separation of amino acids was obtained. It is asserted that resolution (R) of two consecutive chromatographic peaks is complete if, at full width at half height of the smallest one, they do not overlap. Thus, analyzing in more detail the chromatogram of the amino acids in Figure 1, we can notice that complete separation was obtained for all the amino acids except for L-valine and L-β-alanine, L-glutamine and L-glutamic acid, L-histidine and L-homoarginine, with R values of −0.163, −0.059, and −0.054, respectively.

A total of eighteen free amino acids were identified in the Iberian dry-cured ham samples. Figure 2 shows the GC-ion trap-MS chromatogram profile in full scan mode of the trimethyl silyl ethers of the free amino acids fraction isolated according to the method proposed. The tentative assignment of the chromatographic peaks was done comparing the spectra with those from NIST 98 (National Institute of Standards and Technology, Gaithersburg, MD, USA) and WILEY 7 libraries and verified in every single case by standards. Table 2 shows the relative retention times to methimazole, the base peak and the molecular ion for these compounds. The identified amino acids present in Iberian dry-cured ham samples were L-alanine, L-2-aminobutanoic acid, L-aspartic acid, L-cysteine,

L-glutamine, L-glycine, L-histidine, L-hydroxyproline, L-isoleucine, L-leucine, L-lysine, L-methionine, L-phenylalanine, L-proline, L-serine, L-threonine, L-tyrosine, and L-valine. These eighteen amino acids were quantified using methimazole as internal standard, considering that the relative response factor to free amino acids is close to the unit.

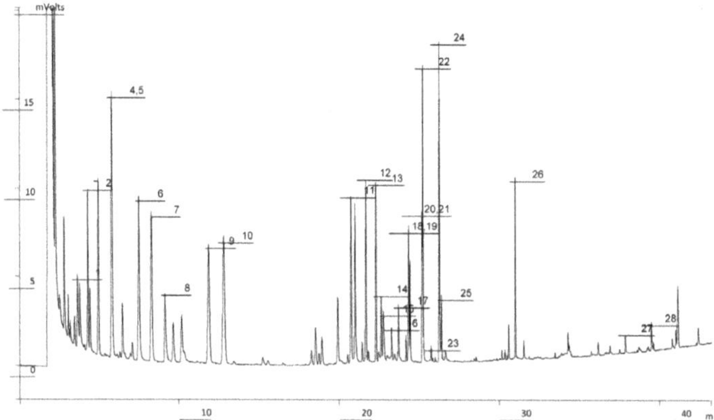

Figure 1. Chromatogram of amino acid standards obtained by GC–FID. Chromatographic conditions are included in paragraph 2.3. Peak numbers corresponding to amino acids appear in Table 1.

Table 1. Amino acids in a standard solution analyzed by GC–FID.

Peak	Amino Acid	T_{rr}
1	L-Alanine	0.167
2	L-Glycine	0.191
3	L-2-Aminobutanoic acid	0.219
4	L-Valine	0.258
5	L-Alanine	0.259
6	L-Leucine	0.332
7	L-Isoleucine	0.369
8	L-Proline	0.429
9	L-Serine	0.533
10	L-Threonine	0.572
11	L-Hydroxyproline	0.934
12	L-Aspartic acid	0.972
13	L-Methionine	1.000
14	L-Arginine	1.009
15	L-Cysteine	1.020
16	L-Citrulline	1.031
17	L-Cystine	1.032
18	L-Glutamine	1.091
19	L-Glutamic acid	1.095
20	L-Histidine	1.127
21	L-Homoarginine	1.130
22	L-Phenylalanine	1.132
23	L-Lysine	1.136
24	L-Asparagine	1.186
25	L-Ornitine	1.203
26	L-Tyrosine	1.373
27	L-Tryptophan	1.659
28	L-Hydroxytryptophan	1.765

Compounds as trimethylsilyl derivatives; T_{rr} relative retention time to L-methionine. See Figure 1 for peak numbers.

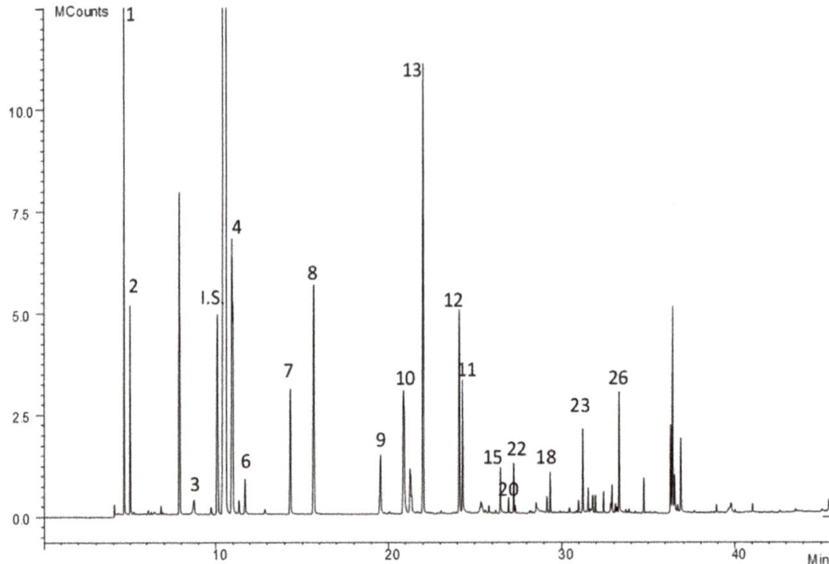

Figure 2. Chromatogram of amino acids in Iberian dry-cured ham obtained by GC–MS. Chromatographic conditions are included in paragraph 2.3. Peak numbers corresponding to amino acids appear in Table 2.

Table 2. Amino acids identified in the Iberian dry-cured ham by GC–MS.

Peak	T_{rr}	Amino Acid	Formula	M$^+$	B.P.
1	0.211	L-Alanine	$C_9H_{23}NO_2Si_2$	233	116
2	0.227	L-Glycine	$C_8H_{21}NO_2Si_2$	219	102
3	0.396	L-2-aminobutanoic acid	$C_{10}H_{25}NO_2Si_2$	247	130
4	0.497	L-Valine	$C_{11}H_{27}NO_2Si_2$	261	144
6	0.532	L-Leucine	$C_{12}H_{29}NO_2Si_2$	275	158
7	0.711	L-Isoleucine	$C_{12}H_{29}NO_2Si_2$	275	158
8	0.650	L-Proline	$C_{11}H_{25}NO_2Si_2$	259	142
9	0.887	L-Serine	$C_{12}H_{31}NO_3Si_3$	321	73
10	0.948	L-Threonine	$C_{13}H_{33}NO_3Si_3$	335	73
13	1.000	L-Methionine	$C_{11}H_{27}NO_2SSi_2$	293	176
12	1.095	L-Aspartic acid	$C_{13}H_{31}NO_4Si_3$	349	73
11	1.103	L-Hydroxyproline	$C_{14}H_{33}NO_3Si_3$	347	229
15	1.202	L-Cysteine	$C_{12}H_{31}NO_2SSi_3$	337	219
22	1.237	L-Phenylalanine	$C_{15}H_{27}NO_2Si_2$	309	73
23	1.417	L-Lysine	$C_{15}H_{38}N_2O_2Si_3$	362	84
18	1.332	L-Glutamine	$C_{14}H_{34}N_2O_3Si_3$	362	156
20	1.495	L-Histidine	$C_{15}H_{33}O_2N_3Si_3$	371	154
26	1.513	L-Tyrosine	$C_{18}H_{35}NO_3Si_3$	397	218

Compounds as trimethylsilyl derivatives; T_{rr}, relative retention time to methimazole; M$^+$, molecular ion; B.P., base peak; See Figure 2 for peak numbers.

Though there are many techniques available for the analysis of amino acids, the previous step of deproteinization is still one of the major problems. Peptides and proteins should be removed because they can interfere in the analysis and separation, as clogging the chromatographic column [33]. Precipitation with 5-sulphosalicylic acid, followed by centrifugation, ultrafiltration, and extraction are some of the most commonly used methods of deproteinization. In meat products, separation with organic solvents like methanol, dichloromethane, and chloroform has been used [23,33]. In this work, a previous step of solvent extraction using n-hexane-diethyl-ether has been performed for degreasing the samples. Then, a mixture of methanol-acetonitrile was used to extract the amino acids.

For each amino acid, the calibration curves were obtained at the corresponding range of linearity. Each curve was prepared six times with a sample in which different amounts of every determined amino acid, at the levels 0, 80%, 100%, and 120%, were added. The calculated equations, area = slope × [mg kg^{-1}] + intercept, are presented in Table 3. As it can be observed, a good correlation was obtained in all cases for a linear fit. The respective peak areas fitted a linear model within the indicate range shown in Table 3. The higher the wideness of linearity range, the more reliable the linear fit is. The slopes of the calibration lines for all amino acids ranged between 314.31 and 730.34, the highest value of this was for L-threonine, and the lowest for L-methionine.

Table 3. Calibration parameters for quantitative determination of amino acids.

Amino Acid	Slope	Intercept	R^2	Linearity Range *
L-Alanine	427.35	−145,386	0.9993	[0.00, 2310.35]
L-Glycine	499.7	−283,535	0.9999	[0.00, 1936.80]
L-2-Aminobutanoic acid	378.51	−22,077	0.9958	[0.00, 683.86]
L-Valine	441.26	−161,050	0.9999	[0.00, 1830.53]
L-Leucine	316.03	43,927	0.9643	[0.00, 1792.65]
L-Proline	346.04	−390.02	0.9969	[0.00, 1426.28]
L-Isoleucine	603.9	−362,852	0.9996	[0.00, 1143.00]
L-Serine	627.99	−159,368	0.9452	[0.00, 543.14]
L-Threonine	730.34	−293,682	0.9993	[0.00, 852.07]
L-Methionine	314.31	588.9	0.9804	[0.00, 35.07]
L-Aspartic acid	422.81	−5239.1	0.9965	[0.00, 66.43]
L-Hydroxyproline	464.86	−23,686	0.9898	[0.00, 209.95]
L-Cysteine	316.75	3075.1	0.9931	[0.00, 1017.84]
L-Phenylalanine	353.7	36.418	0.9998	[0.00, 434.64]
L-Glutamine	445.06	−38,852	0.9999	[0.00, 865.38]
L-Lysine	347.24	−5227.5	0.9840	[0.00, 578.92]
L-Histidine	358.7	−457.87	0.9991	[0.00, 606.50]
L-Tyrosine	343.04	−136.99	0.9995	[0.00, 532.76]

* mg kg^{-1}.

The trueness was assessed based on recovery assays that were carried out in the following way. A sample of Iberian dry-cured ham muscle was analyzed ten times by the proposed method. The obtained results for the different amino acids have been used as reference values. Then, different amounts of every determined amino acid at 80%, 100%, and 120% levels were added to the same sample, and ten replicates have been done for each case. The obtained results are shown in Table 4. As it can be seen, recoveries lie within the range 52–164% that can be considered acceptable values according the analyzed concentration [34] and consequently, trueness is significant.

For the determination of repeatability, replicates were done on different days and in the same laboratory. Table 4 shows the obtained results. Relative standard deviations (RSD) range between 6.28% and 17.81%. These values of RSD are minor compared with the reference value derived from Horwitz equation. Therefore, the results for different amino acids indicate a good repeatability for the assay.

The limits of detection (LOD) and limits of quantitation (LOQ) were obtained. The LOD of the method was determined considering a signal-to-noise ratio of 3 with reference to the background noise obtained from a blank sample. LOQ was calculated considering a signal-to-noise ratio of 10. Values of LOD and LOQ are shown in Table 4. The LOD obtained were between 7.0 and 611.74 mg kg^{-1}, and LOQ ranged from 27.9 to 1779.9 mg kg^{-1}. The lowest LOD and LOQ were for L-methionine, and the highest for L-isoleucine.

Table 4. Repeatability, recovery, LOD (limits of detection), and LOQ (limits of quantitation) of amino acids analysis in Iberian dry-cured ham by GC–FID.

Amino Acid	Mean *	SD	RSD	Recovery	LOD *	LOQ *
L-Alanine	2336.9	151.8	6.5	101.36	457.2	1468.0
L-Glycine	1139.8	104.0	9.1	102.63	516.2	1521.0
L-2-aminobutanoic acid	358.0	59.6	16.7	113.05	55.5	160.3
L-Valine	920.2	80.4	8.7	121.31	322.7	913.9
L-Leucine	730.6	84.7	11.6	144.12	172.5	641.9
L-Proline	836.8	52.5	6.3	107.73	611.7	1779.9
L-Isoleucine	514.9	55.4	10.8	133.92	79.2	261.3
L-Serine	332.6	42.4	12.8	94.54	250.6	732.2
L-Threonine	417.9	39.0	9.3	113.02	308.5	1067.3
L-Methionine	17.1	3.0	17.5	164.13	7.0	27.9
L-Aspartic acid	64.1	11.2	17.4	52.29	32.0	77.8
L-Hydroxyproline	136.5	13.0	9.5	99.10	100.5	303.3
L-Cysteine	770.9	137.3	17.8	77.68	427.4	1484.7
L-Phenylalanine	158.4	19.3	12.2	157.30	96.9	262.1
L-Glutamine	575.0	38.7	6.7	91.34	362.2	1219.6
L-Lysine	461.0	29.5	6.4	77.83	231.1	735.3
L-Histidine	291.5	33.9	11.6	114.73	17.2	54.4
L-Tyrosine	258.3	12.7	4.9	131.43	196.1	845.4

* mg kg^{-1}; average of the three spiking levels (80%, 100%, 120%); SD, standard deviation; RSD, relative standard deviation, RSD and Recovery in %.

Figure 3 shows a GC–FID chromatogram of a sample of dry-cured ham. At the beginning of this amino acids chromatogram, there are some high peaks corresponding to free fatty acids, according to previously described in literature [26]. As it is shown in Figure 3, the free amino acids in Iberian ham muscle samples in higher amounts are L-alanine, L-glycine, L-valine, and L-proline, followed by L-leucine, L-cysteine, L-glutamine, and L-isoleucine. However, this fact disagrees with literature from other authors using a different technique [35], in which the Iberian ham, in its ripening end step, had higher amounts of the free amino acids L-glutamic acid, L-alanine, L-leucine, and L-glycine [35]. On the other hand, free amino acids in lesser amount were L-aspartic acid and L-methionine. Moreover, L-glutamic acid had not been detected [35].

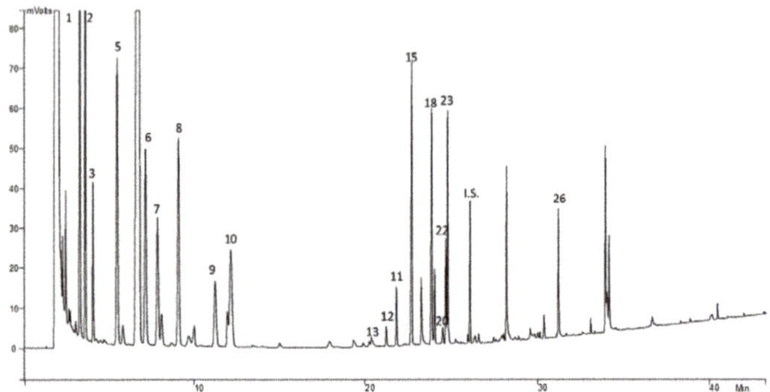

Figure 3. Chromatogram of amino acids in Iberian dry-cured ham obtained by GC–FID. Chromatographic conditions are included in Setcion 2.2. Peak numbers corresponding to amino acids appear in Table 2.

3.2. Evolution of Free Amino Acids during Dry-Curing Process

Concentrations of amino acids in samples of Iberian dry-cured ham at different times of the ripening process have been determined. Samples at 1, 33, 64, 99, 180, 245, 309, 363, and 435 days were taken and analyzed according to the proposed GC–FID method. The sampling days account for the three considered ripening stages, postsalting, drying, and cellar. As it has been mentioned above, eighteen free amino acids have been detected in the samples throughout the ripening period of dry-cured Iberian ham. Results are included in Table 5. The total amount of free amino acids all along the process increases, which agrees with previous results described in the literature [35]. From a starting value of 6387.32 mg Kg^{-1} to a final one of 10320.55 mg kg^{-1}, the following equation can be considered: [amino acid] = 5.7767 × (days) + 4714.6; R^2 = 0.8034. Table 6 includes the variation of percentages of free amino acids according to the ripening time. In this way, an easier form to visualize the evolution of the amino acids in the samples is achieved. Figure 4 shows these percentages grouped for the different amino acids. As it can be seen in this figure, L-alanine is the main amino acid, with a significant decrease during postsalting and drying process, showing a maximum value of 59.56% of the total free amino acids fraction at the beginning of the process, and it stabilizes around 25.00% all along the cellar step (Figure 4(1)). The other main amino acid found in samples is L-cysteine, which remains fixed all along the process between 23.77% and 20.31%, as shown in Figure 4(1). L-lysine (Figure 4(2)), L-valine, L-leucine, and L-isoleucine (Figure 4(3)) increase all along the ripening process with a linear trend. On the other hand, the remainder components of this fraction suffer an increase that may be considered linear during postsalting and drying process to be fixed in cellar stage (Figure 4(3–6)). Only L-aspartic acid does not show, apparently, any trend during all the process.

Table 5. Variation of concentration * of amino acids according to the ripening time.

				Ripening Time (Days)					
Amino acid	1	33	64	99	180	245	309	363	435
L-Alanine	12,175.0	10,973.1	8392.8	6015.3	8610.5	5011.0	6349.8	4631.1	10,409.5
L-Glycine	512.8	922.9	914.9	967.2	2695.5	87.7	108.9	89.4	126.4
L-2-aminobutanoic acid	127.4	163.2	234.4	225.7	909.8	1655.6	1931.1	1115.2	2990.7
L-Valine	312.1	468.8	530.4	642.0	2262.4	471.9	528.5	308.8	957.9
L-Leucine	307.1	493.2	509.7	701.1	2748.9	1709.1	1890.9	957.5	2814.1
L-Isoleucine	166.0	272.3	290.4	400.3	1586.8	2164.6	2295.7	1035.5	3207.1
L-Proline	239.5	397.6	382.5	446.7	1657.8	1212.9	1379.1	607.9	1953.3
L-Serine	74.8	212.1	196.5	196.7	990.8	1142.1	1438.7	657.4	2014.2
L-Threonine	218.2	252.1	225.9	274.0	735.2	495.6	720.9	342.5	1086.8
L-Methionine	23.6	54.7	45.2	82.7	448.2	154.5	479.5	264.2	852.0
L-Aspartic acid	4545.8	4270.4	4317.4	4175.1	6602.8	296.9	271.3	97.5	360.3
L-Hydroxyproline	66.9	57.6	39.5	82.7	396.3	3650.9	6446.7	4515.1	8892.4
L-Cysteine	171.8	198.8	207.0	352.6	1264.5	691.4	799.8	450.0	1490.0
L-Phenylalanine	65.1	87.6	114.2	147.0	804.0	345.3	390.1	205.2	684.8
L-Lysine	44.2	44.6	38.8	77.3	204.3	203.9	42.0	53.8	357.8
L-Glutamine	155.9	256.7	251.9	340.6	1330.8	1027.9	1242.7	512.2	1510.4
L-Histidine	1448.6	1252.7	1376.7	1287.1	3420.0	1743.4	1622.3	1285.5	2645.8
L-Tyrosine	122.7	70.5	51.1	54.4	117.2	154.1	60.3	29.8	90.2
TOTAL	20,778.5	20,481.9	18,183.4	16,567.3	36,965.8	22,463.9	28,307.2	17,521.6	42,878.7

* mg kg^{-1}.

Table 6. Variation of percentages (%) of amino acids according to the ripening time.

				Ripening Time (Days)					
Amino acid	1	33	64	99	180	245	309	363	435
L-Alanine	58.60	53.66	46.32	36.53	23.41	22.55	22.68	26.99	24.53
L-Glycine	2.47	4.51	5.05	5.87	7.33	0.39	0.39	0.52	0.30
L-2-aminobutanoic acid	0.61	0.80	1.29	1.37	2.47	7.45	6.90	6.50	7.05
L-Valine	1.50	2.29	2.93	3.90	6.15	2.12	1.89	1.80	2.26
L-Leucine	1.48	2.41	2.81	4.26	7.47	7.69	6.75	5.58	6.63
L-Isoleucine	0.80	1.33	1.60	2.43	4.31	9.74	8.20	6.03	7.56
L-Proline	1.15	1.94	2.11	2.71	4.51	5.46	4.93	3.54	4.60
L-Serine	0.36	1.04	1.08	1.19	2.69	5.14	5.14	3.83	4.75
L-Threonine	1.05	1.23	1.25	1.66	2.00	2.23	2.57	2.00	2.56
L-Methionine	0.11	0.27	0.25	0.50	1.22	0.70	1.71	1.54	2.01
L-Aspartic acid	21.88	20.88	23.83	25.35	17.95	1.34	0.97	0.57	0.85
L-Hydroxyproline	0.32	0.28	0.22	0.50	1.08	16.43	23.03	26.31	20.95
L-Cysteine	0.83	0.97	1.14	2.14	3.44	3.11	2.86	2.62	3.51
L-Phenylalanine	0.31	0.43	0.63	0.89	2.19	1.55	1.39	1.20	1.61
L-Lysine	0.21	0.22	0.21	0.47	0.56	0.92	0.15	0.31	0.84
L-Glutamine	0.75	1.26	1.39	2.07	3.62	4.63	4.44	2.99	3.56
L-Histidine	6.97	6.13	7.60	7.82	9.30	7.85	5.79	7.49	6.23
L-Tyrosine	0.59	0.34	0.28	0.33	0.32	0.69	0.22	0.17	0.21

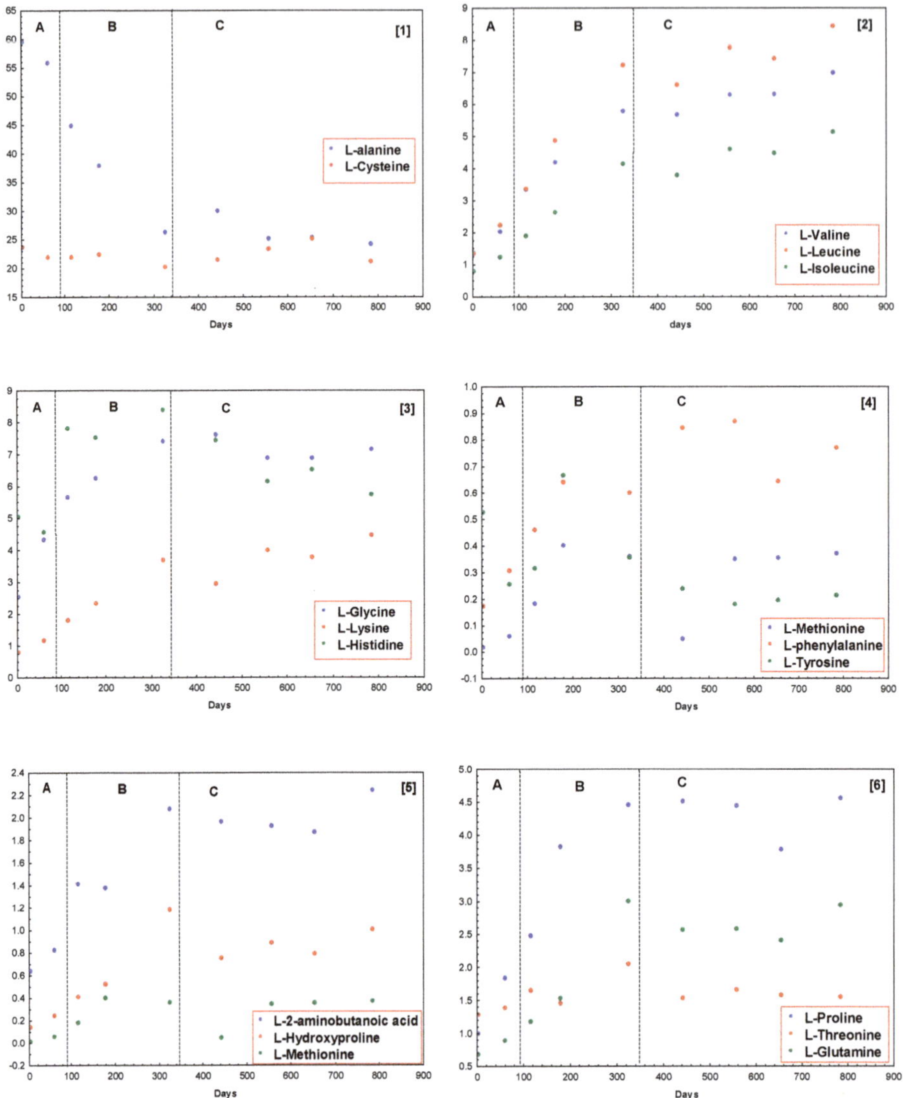

Figure 4. Variation of percentages of different amino acids (from 1 to 6) in Iberian dry-cured ham according to the ripening time.

3.3. Differentiation between Ripening Stages

To differentiate between the three ripening stages considered, postsalting, drying, and cellar, some chemometric calculations have been performed using the percentages of free amino acids obtained in the analysis of the samples. By applying PCA to the data matrix, two PCs were obtained. PC1 explains 69.08% of the variance, and PC2 9.66%, accounting these two PCs for 78.64% of the total variance. Figure 5A shows the scores plot considering the two first PCs. PC1 is strongly influenced by L-alanine, L-glycine, L-2-aminobutanoic acid, L-valine, L-leucine, L-isoleucine, L-proline, L-serine, L-aspartic acid, L-methionine, L-glutamine, L-phenylalanine, and L-lisine, and PC2 by L-cysteine, and L-histidine. The correlation between these variables is higher than 0.7. Though the samples corresponding to the different stages are not completely separated, some tendencies can be appreciated. Samples at postsalting stage are located in the scores plot at positive values of PC1, and those corresponding to cellar stage are situated at negative values of PC1. However, the samples corresponding to drying period appear spread between positive and negative values of PC1. An overlapping can be appreciated, and it could be explained considering this is a continuous process, being the first samples of drying mixed with those of postsalting and the last samples of drying mixed with the samples of cellar. This is an important consideration because in the process followed in processing plants, there are no clear criteria for transferring hams from drying to cellar stages, and the results obtained in this study can be very useful to optimize the elaboration process.

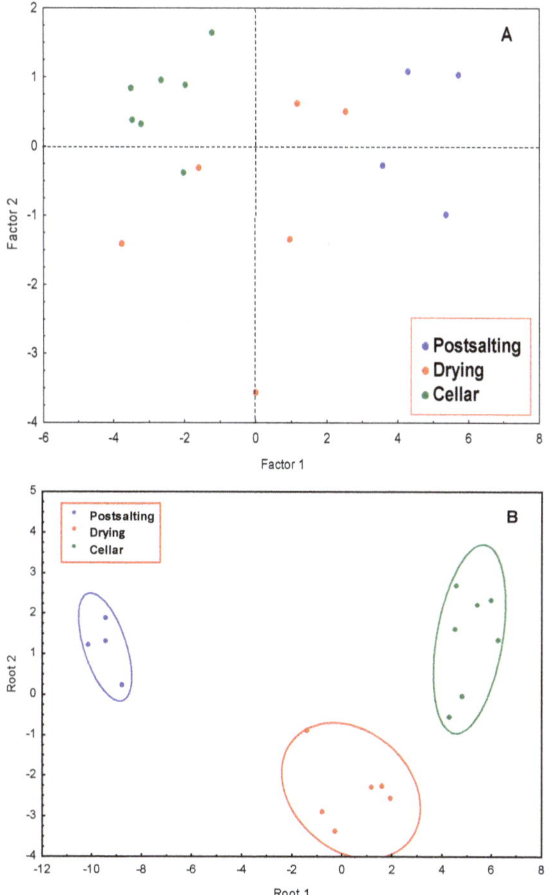

Figure 5. (**A**) Scores plot of Iberian dry-cured ham samples for the first PCs; blue, red, and green dots correspond to postsalting, drying, and cellar samples, respectively; (**B**) Scatter plot of the Iberian dry-cured ham samples in the plane of the two first DFs (discriminant functions).

A classification rule to differentiate between the three stages was obtained by applying LDA. Using Wilks' Lambda of 0.00532, two discriminant functions (Root 1 and 2) were obtained. Backward stepwise analysis retained the amino acids L-alanine, L-tyrosine, L-glutamine, L-proline, L-2-aminobutanoic acid, L-cysteine, and L-valine that can be considered the most discriminant variables. The classification functions are shown in Table 7. A complete separation between the three stages was obtained, denoting that the considered variables are powerful descriptors to differentiate between the samples from three dry-curing periods considered, postsalting, drying, and cellar.

Table 7. Results of the stepwise LDA.

Amino Acid	Postsalting [#]	Drying [#]	Cellar [#]	F to Remove	*p*-Level
L-alanine	269.1	275.5	283.3	5.42201	0.032484
L-Tyrosine	251.4	265.3	271.1	4.42997	0.050692
L-Glutamine	−128.0	−173.5	−179.8	6.02760	0.025319
L-Proline	664.1	703.9	723.3	8.42128	0.010754
L-2-Aminobutanoic acid	1306.4	1398.7	1440.2	7.51229	0.014574
L-Cysteine	359.8	374.1	385.4	10.85224	0.005261
L-Valine	1410.9	1448.3	1493.2	7.12056	0.016739
Constant	−14584.7	−15553.3	−16467.7		

[#] $p = 0.3333$; *F* value, 2.8.

4. Conclusions

By using GC–MS and GC–FID, eighteen free amino acids have been identified and determined in Iberian dry-cured ham at different ripening stages. Postsalting, drying, and cellar stages have been considered. During the ripening process, significant increasing of the amounts of all free amino acids, except L-alanine, was appreciated. To differentiate between the three ripening stages and using the free amino acids as chemical descriptors, PCA and LDA were applied. Using LDA, a total differentiation between the three ripening periods was obtained. The trends of total free amino acids in postsalting, drying, and cellar periods, respectively, can be used in order to predict the curing time.

Author Contributions: All authors have made the same contribution to the article. All authors have read and agree with the published version of the manuscript.

Funding: This study was supported by Junta de Andalucía through the project P09-AGR-04789.

Acknowledgments: The authors are grateful especially to J.L. León-García for the collaboration and given help.

Conflicts of Interest: The authors declare no conflict of interest.

References

1. Gallardo, E.; Narváez-Rivas, M.; Pablos, F.; Jurado, J.M.; León-Camacho, M. Subcutaneous fat triacylglycerols profile from Iberian pigs as a tool to differentiate between intensive and extensive fatening systems. *J. Agric. Food Chem.* **2012**, *60*, 1645–1651. [CrossRef] [PubMed]

2. Bénédicte, L.; Guillard, A. Outdoor rearing of cull sows: Effects on carcass, tissue and meat quality. *Meat Sci.* **2005**, *70*, 247–257.

3. Rey, A.L.; Daza, A.; López-Carrasco, C.; López-Bote, C.J. Feeding Iberian pigs with acorns and grass in either free-range or confinement affects the carcass characteristics and fatty acids and tocopherols accumulation in Longissimus dorsi muscle and backfat. *Meat Sci.* **2006**, *73*, 66–74. [CrossRef] [PubMed]

4. Regulation (EU) No 775/2010. *Entering a Name in the Register of Protected Designations of Origin and Protected Geographical Indications (Los Pedroches (PDO))*; EU: Brussels, Belgium, 2010.

5. Wang, Z.Y.; Gao, X.G.; Zhang, J.H.; Zhang, D.Q.; Ma, C.W. Changes of intramuscular fat composition, lipid oxidation and lipase activity in biceps femoris and semimembranous of Xuanwei ham during controlled salting stages. *J. Integr. Agric.* **2013**, *12*, 1993–2001. [CrossRef]

6. Kato, H.; Rhue, M.R.; Nishimura, T. Role of free amino acids and peptides in food taste. In *Flavor Chemistry*; Teranishi, R., Buttery, R.G., Shahidi, F., Eds.; ACS Publications: Washington, DC, USA, 1989; Volume 388, pp. 158–174.

7. Petrova, I.; Aasen, I.M.; Rustad, T.; Eikevik, T.M. Manufacture of dry-cured ham: A review. Part 1. Biochemical changes during the technological process. *Eur. Food Res. Technol.* **2015**, *241*, 587–599. [CrossRef]

8. Toldra, F. *Dry-Cured Meat Products*; Food and Nutrition Press: Trumbull, CT, USA, 2002; p. 157.

9. Narváez-Rivas, M.; Vicario, I.M.; Constante, E.G.; León-Camacho, M. Changes in the fatty acid and triacylglycerol profiles in the subcutaneous fat of Iberian ham during the dry-curing process. *J. Agric. Food Chem.* **2008**, *56*, 7131–7137. [CrossRef]

10. Narváez-Rivas, M.; Gallardo, E.; León-Camacho, M. Changes in polar and non-polar lipid fractions of subcutaneous fat from Iberian ham during dry-curing process. Prediction of the curing time. *Food Res. Int.* **2013**, *54*, 213–222. [CrossRef]

11. Narváez-Rivas, M.; Gallardo, E.; León-Camacho, M. Chemical changes in volatile aldehydes and ketones from subcutaneous fat during ripening of Iberian dry-cured ham. Prediction of the curing time. *Food Res. Int.* **2014**, *55*, 381–390. [CrossRef]

12. Narváez-Rivas, M.; Gallardo, E.; León-Camacho, M. Evolution of volatile hydrocarbons from subcutaneous fat during ripening of Iberian dry-cured ham. A tool to differentiate between ripening periods of the process. *Food Res. Int.* **2015**, *67*, 299–307. [CrossRef]

13. AOAC. *Official Methods of Analysis of AOAC International*, 17th ed.; AOAC: Gaithesrsburg, MD, USA, 2002; Volume 1, Chapter 4; p. 6C.

14. Molnár-Perl, J. Role of chromatography in the analysis of sugars, carboxylic acids and amino acids in food. *J. Chromatogr. A* **2000**, *891*, 1–32. [CrossRef]

15. Callejón, R.M.; Troncoso, A.M.; Morales, M.L. Determination of amino acids in grape-derived products: A review. *Talanta* **2010**, *81*, 1143–1152. [CrossRef] [PubMed]

16. Konstantinos Petritis, K.; Elfakir, C.; Dreux, M. A comparative study of commercial liquid chromatography detectors for the analysis of underivatized amino acids. *J. Chromatogr. A* **2002**, *961*, 9–21. [CrossRef]

17. Gökmen, V.; Serpen, A.; Mogol, B.A. Rapid determination of amino acids in foods by hydrophilic interaction liquid chromatography coupled to high–resolution mass spectrometry. *Anal. Bioanal. Chem.* **2012**, *403*, 2915–2922. [CrossRef] [PubMed]

18. Kataok, H.; Matsumura, S.; Koizumi, H.; Makita, M. Rapid and simultaneous analysis of protein and non-protein amino acids as N(O,S)-isobutoxycarbonyl methyl ester derivatives by capillary gas chromatography. *J. Chromatogr. A* **1997**, *758*, 167–173. [CrossRef]

19. Cao, P.; Moini, M. Quantitative analysis of fluorinated ethylchloroformate derivatives of protein amino acids and hydrolysis products of small peptides using chemical ionization gas chromatography-mass spectrometry. *J. Chromatogr. A* **1997**, *759*, 111–117. [CrossRef]

20. Haberhauer-Troyer, C.; Álvarez-Llamas, G.; Zitting, E.; Rodríguez-González, P.; Rosenberg, E.; Sanz-Medel, A. Comparison of different chloroformates for the derivatisation of seleno amino acids for gas chromatographic analysis. *J. Chromatogr. A* **2003**, *1015*, 1–10. [CrossRef]

21. Dallakian, P.; Budzikiewicz, H. Gas chromatography–chemical ionization mass spectrometry in amino acid analysis of pyoverdins. *J. Chromatogr. A* **1997**, *787*, 195–203. [CrossRef]

22. Chiavari, G.; Fabbri, D.; Prati, S. Gas chromatographic–mass spectrometric analysis of products arising from pyrolysis of amino acids in the presence of hexamethyldisilazane. *J. Chromatogr. A* **2001**, *922*, 235–241. [CrossRef]

23. Leggio, A.; Belsito, E.L.; De Marco, R.; Liguori, A.; Siciliano, C.; Spinella, M. Simultaneous extraction and derivatization of amino acids and free fatty acids in meat products. *J. Chromatogr. A* **2012**, *1241*, 96–102. [CrossRef]

24. Nozal, M.J.; Bernal, J.L.; Toribio, M.L.; Diego, J.C.; Ruíz, A. Rapid and sensitive method for determining free amino acids in honey by gas chromatography with flame ionization or mass spectrometric detection. *J. Chromatogr. A* **2004**, *1047*, 137–146. [CrossRef]

25. Husek, P. Rapid derivatization and gas chromatographic determination of amino acids. *J. Chromatogr. A* **1991**, *552*, 289–299. [CrossRef]

26. Paik, M.J.; Kim, K.R. Sequential ethoxycarbonylation, methoximation and tert- butyldimethylsilylation for simultaneous determination of amino acids and carboxylic acids by dual-column gas chromatography. *J. Chromatogr. A* **2004**, *1034*, 13–23. [CrossRef] [PubMed]

27. Wold, S.; Esbensen, K.; Geladi, P. Principal Component Analysis. *Chemom. Intell. Lab. Syst.* **1987**, *2*, 37–52. [CrossRef]

28. Chatfield, C.; Collins, A. Principal Component Analysis. In *Introduction to Multivariate Analysis*; Chapman & Hall: London, UK, 1980; pp. 57–81.

29. Gardiner, W.P. *Statistical Analysis Methods for Chemists*; Royal Society of Chemistry: Cambridge, UK, 1997; pp. 167–181.

30. Danezis, G.P.; Tsagkaris, A.S.; Camin, F.; Brusic, V. Food authentication: Techniques, trends & emerging approaches. *Trends Anal. Chem.* **2016**, *85*, 123–132.

31. Chaves Das Neves, H.J.C.; Vasconcelos, A.M.P. Capillary gas chromatography of amino acids, including asparagine and glutamine: Sensitive gas chromatogaphic-mass spectrometric and selected ion monitoring gas chromatographic-mass spectrometric detection of the N.,O(S)-tert.-butyldimethylsilyl derivative. *J. Chromatogr. A* **1987**, *392*, 249–258. [CrossRef]
32. Dauner, M.; Sauer, U. GC–MS analysis of amino acids rapidly provides rich information for isotopomer balancing. *Biotechnol. Prog.* **2000**, *16*, 642–649. [CrossRef]
33. Liu, H.; Worthen, H.G. Measurement of free amino acid levels in ultrafiltrates of blood plasma by high-performance liquid chromatography with automatic pre-column derivatization. *J. Chromatogr. B Biomed. Sci. Appl.* **1992**, *579*, 215–224. [CrossRef]
34. González, A.G.; Herrador, M.A. A practical guide to analytical method validation, including measurement uncertainty and accuracy profiles. *Trends Anal. Chem.* **2007**, *26*, 227–238. [CrossRef]
35. Córdoba, J.J.; Rojas, T.A.; González, C.G.; Barroso, J.V.; Bote, C.L.; Asensio, M.A. Evolution of Free Amino Acids and Amines during Ripening of Iberian Cured Ham. *J. Agric. Food Chem.* **1994**, *42*, 2296–2301. [CrossRef]

 *foods*

Article

Multi-Chemical Profiling of Strawberry as a Traceability Tool to Investigate the Effect of Cultivar and Cultivation Conditions

Raúl González-Domínguez [1,2,*], Ana Sayago [1,2], Ikram Akhatou [1,2] and Ángeles Fernández-Recamales [1,2,*]

[1] Department of Chemistry, Faculty of Experimental Sciences, University of Huelva, 21007 Huelva, Spain; ana.sayago@dqcm.uhu.es (A.S.); ikram.akhatou@alu.uhu.es (I.A.)

[2] International Campus of Excellence CeiA3, University of Huelva, 21007 Huelva, Spain

* Correspondence: raul.gonzalez@dqcm.uhu.es (R.G.-D.); recamale@dqcm.uhu.es (Á.F.-R.); Tel.: +34-959219975 (R.G.-D.); +34-959-219958 (Á.F.-R.)

Received: 23 December 2019; Accepted: 13 January 2020; Published: 16 January 2020

Abstract: The chemical composition of foods is tightly regulated by multiple genotypic and agronomic factors, which can thus serve as potential descriptors for traceability and authentication purposes. In the present work, we performed a multi-chemical characterization of strawberry fruits from five varieties (Aromas, Camarosa, Diamante, Medina, and Ventana) grown in two cultivation systems (open/closed soilless systems) during two consecutive campaigns with different climatic conditions (rainfall and temperature). For this purpose, we analyzed multiple components closely related to the sensory and health characteristics of strawberry, including sugars, organic acids, phenolic compounds, and essential and non-essential mineral elements, and various complementary statistical approaches were applied for selecting chemical descriptors of cultivar and agronomic conditions. Anthocyanins, phenolic acids, sucrose, and malic acid were found to be the most discriminant variables among cultivars, while climatic conditions and the cultivation system were behind changes in polyphenol contents. These results thus demonstrate the utility of combining multi-chemical profiling approaches with advanced chemometric tools in food traceability research.

Keywords: strawberry; traceability; sugars; organic acids; phenolic compounds; mineral elements; cultivar; cultivation system

---

1. Introduction

The composition of foods, in terms of nutrients, bioactive compounds, and other components, is tightly regulated by multiple factors, such as the genotype, geographical origin, environmental factors, and agronomic conditions. Therefore, this influences the sensory, nutritional, and nutraceutical properties of food products, which makes the implementation of quality control strategies mandatory to ensure their authenticity and traceability. In this vein, it should be noted that food quality and safety may be influenced by a myriad of factors throughout the entire supply chain, from initial food production to packaging, processing, and transport, until its final commercialization [1]. This is particularly important for processed foods, which usually require more complex operations and thus make the implementation of efficient traceability initiatives mandatory. To address these needs, novel and powerful analytical methods are requested by the food industry to accurately guarantee the authenticity and traceability of food products.

Strawberry (*Fragaria* × *ananassa* Duch.) is one of the most commonly consumed berry fruits around the world and is considered a functional food because of its chemical composition, which rich in essential and bioactive compounds. Strawberry has been demonstrated to lower post-prandial oxidative stress, hyperglycemia, hyperlipidemia, and inflammation, and its consumption has been associated with a reduced incidence of cardiovascular diseases (e.g., hypertension), cancer, and other diseases [2,3]. In this context, this berry fruit has been proposed as a potential ingredient for the production of nutraceutical products, such as beverages, flours, and powders [4]. The main soluble constituents of strawberry include sugars (e.g., glucose, fructose, and sucrose) and organic acids (e.g., citric and malic), which influence the final taste and flavor of this fruit [5]. Furthermore, strawberry is also a rich source of numerous bioactive compounds, such as dietary fibers, minerals, vitamins, and phenolic compounds [6,7]. In particular, polyphenols are known to elicit multiple biological activities, acting as natural antioxidants that protect the organism against free radicals [8]. Other pro-healthy compounds found in strawberries are vitamins and minerals, which intervene in a multitude of processes and chemical reactions inside the cells. For instance, potassium, a major element in strawberry, plays an important role in protection against cardiovascular diseases [9]. According to recent literature, this characteristic chemical profile of strawberry is largely influenced by multiple factors (e.g., cultivar, climate, and cultivation conditions) [7,10–13], evidencing its potential as a traceability tool.

In this work, we employed a multi-targeted profiling approach to characterize the chemical composition of strawberry, considering multiple compounds related to sensory and health characteristics of this berry fruit, including sugars, organic acids, polyphenols, and mineral elements. This multi-chemical profile was investigated as a potential tool for authentication and traceability purposes, with the aim of discriminating strawberry varieties grown under different climatic and agronomic conditions. For this purpose, complementary pattern recognition procedures were employed, including principal component analysis (PCA), linear discriminant analysis (LDA), soft independent model class analogy (SIMCA), and partial least squares discriminant analysis (PLS-DA).

2. Materials and Methods

2.1. Experimental Design and Sampling

Strawberry fruits (*Fragaria* × *ananassa* Duch.) were collected in two consecutive campaigns (years 2015 and 2016) from the same experimental plantations located in Huelva (southwest Spain), at the same commercial ripeness (>75% of the surface showing red color). The first campaign was characterized by higher total radiation, while in the second one, higher rainfall, and maximum and minimum temperatures were registered. Five varieties of strawberries, genetically characterized by the vendor (Aromas, Camarosa, Diamante, Medina, and Ventana) and grown in two soilless systems (closed and open systems, i.e., with and without recirculation of the nutrient solution, respectively), were investigated. Plants were grown in a polycarbonate-covered greenhouse using elevated horizontal troughs filled with coconut fiber as a substrate, and with natural daylight as a radiation source. The temperature ranged from 25 °C during the day to 8 °C at night, with relative humidity held at 75 ± 5%.

Several fruits (n = 10) were collected for each variety and cultivation system to generate a representative pooled sample. Immediately after harvesting, fruits were sorted, frozen in situ in a deep freezer, and shipped to the laboratory in polystyrene punnets. Then, fruits were washed, sepals were dissected, and pooled fruits (n = 10) were gently homogenized by using a kitchen mixer to obtain a puree (approximately 100–150 mL). Samples were subsequently aliquoted and stored for up to 2 months at −21 °C, until further analysis. For each study condition (i.e., cultivar, campaign, and cultivation conditions), three replicates (i.e., three pooled and homogenized samples) were prepared.

2.2. Analysis of Sugars and Organic Acids

Sugars and organic acids were analyzed using an Agilent 110 series high-performance liquid chromatography (HPLC) system coupled to ultraviolet (UV) and refractive index (RI) detectors (Agilent Technologies, Santa Clara, CA, USA), following the methodology previously described [7]. Approximately 1 g of the homogenate was accurately weighed, diluted to 10 mL with ultrapure water

(Millipore, Bedford, Massachusetts, MA, USA), and centrifuged at 10,000 rpm for 10 min (BHG-Hermle Z 365, Wehingen, Germany). The supernatant was filtered through a 0.45 μm PVDF (polyvinylidene difluoride) filter prior to HPLC analysis.

In a single chromatographic run, three sugars (glucose, fructose, and sucrose) and six organic acids (oxalic, citric, tartaric, malic, succinic, and lactic) were separated using a Metacarb 87H hydrogen-form cation-exchange resin-based column (300 × 7.8 mm internal diameter, i.d.) packed with sulfonated polystyrene. A total of 5 mM of sulfuric acid was delivered in isocratic mode at a 0.5 mL min^{-1} flow rate for 15 min, and the injection volume was 20 μL. UV detection of organic acids was performed at 210 nm, while sugars were analyzed by using the RI detector. Identifications were accomplished by comparing retention times (and UV spectra for organic acids) with those of reference standards.

2.3. Analysis of Phenolic Compounds

Homogenized fruits (5.0 g) were dissolved with 25 mL of methanol, sonicated for 30 min, and then centrifuged at 10,000 rpm for 10 min at 4 °C. Supernatants were concentrated by using a rotary evaporator at 40 °C, and the residues were re-dissolved in 3 mL of 50% methanol (*v/v*). The concentrated extracts were filtered through 0.45 μm PVDF filters, and 20 μL was injected into a reverse phase Ultrabase C18 column (2.5 μm, 100 mm × 4.6 mm i.d.), following the methodology described elsewhere [10]. For the analysis of colorless flavonoids and phenolic acids, elution solvents were water:methanol:acetic acid (93:5:2, *v/v/v*) (eluent A) and methanol:acetic acid (98:2, *v/v*) (eluent B), which were delivered as follows: 0–29 min, 40% B; 29–34.8 min, 40–60% B; 34.8–37.7 min, 60–75% B; 37.7–40.6 min, 75–100% B; 40.6–46.4 min, 0% B. For anthocyanins, the mobile phase consisted of 10% (*v/v*) aqueous formic acid (eluent A) and methanol (eluent B), using the following gradient program: 0–0.70 min, 5% B; 0.70–16.60 min, 5–50% B; 16.60–18.60 min, 50–95% B; 18.60–20.60 min, 95% B. The flow rate was 0.8 mL min^{-1}, and the column temperature was set at 20 and 30 °C for non-anthocyanin and anthocyanin compounds, respectively.

The identification of phenolic compounds was achieved by comparing their retention times and UV spectra with those for commercial standards. For quantification, the following wavelengths were employed: 260 nm for ellagic acid and derivatives, 280 nm for benzoic acids and flavan-3-ols, 320 nm for cinnamic acids, 360 nm for flavonols, and 520 nm for anthocyanins.

2.4. Analysis of Mineral Elements

For mineral content analysis, 0.5 g of fruit was placed in a Teflon vessel and digested with 3 mL of a mixture of nitric and hydrochloric acids, both 1.5 M. Digestion was carried out for 2 min using a microwave furnace at 250 W. After cooling, the digest was filtered, transferred to a 25 mL flask, and made-up with ultrapure water.

The major (i.e., Ca, Mg, K, P, and S) and trace (i.e., Al, As, Cd, Cr, Cu, Fe, Hg, Ni, Pb, Ba, Mn, Na, Sr, V, and Zn) minerals were determined by inductively coupled plasma optical emission spectrometry (ICP-OES) using a Jobin-Yvon Ultima 2 ICP spectrometer with an ultrasonic nebulizer (U6000 AT+, Cetac). The instrument was operated at the following conditions: radio frequency, 27 MHz; operating power, 1200 W; plasma argon flow rate, 2 L min^{-1}; auxiliary gas flow rate, 2 L min^{-1}; nebulizer gas flow rate, 0.02 L min^{-1}; nebulizer pressure, 1 bar; rinsing time, 35 s; rinsing pump speed, high; transfer time, 60 s; stabilization time, 20 s; and transfer pump speed, high. ICP Multi Element Standard IV and VI CertiPur® (Merck) were used to prepare reference solutions.

2.5. Statistical Analysis

One-way analysis of variance (ANOVA), multivariate analysis of variance (MANOVA), and pattern recognition techniques, including principal component analysis (PCA), linear discriminant analysis (LDA), soft independent modeling of class analogy (SIMCA), and partial least squares discriminant analysis (PLS-DA), were carried out to investigate the differences among strawberry varieties and/or cultivation systems. All statistical analyses were conducted on Statistica 7.1 (StatSoft Inc., Tulsa, Oklahoma, OK, USA) and SIMCA-P™ 11.5 (UMetrics AB, Umeå, Sweden).

3. Results and Discussion

3.1. Multi-Chemical Profiling of Strawberry

Mean concentrations for all the analyzed compounds (i.e., sugars, organic acids, polyphenols, and mineral elements) are listed in Table 1 for the five strawberry cultivars investigated. Soluble sugars identified and quantified in strawberry fruits were fructose, glucose, and sucrose; monosaccharides were the major species in all varieties, except for "Camarosa", which showed higher sucrose contents. The ratio of fructose to glucose content was about the same, regardless of the cultivar, in agreement with our previous study findings [10]. With regards to organic acids, citric acid was the most concentrated metabolite, followed by malic acid, in consonance with previous studies [7,11]. In agreement with results found in the literature, anthocyanins were the predominant polyphenol class in strawberry [14], followed by phenolic acids, with pelargonidin 3-glucoside, pelargonidin 3-rutinoside, and cyanidin 3-glucoside being the three major anthocyanin species [8,15], which were found at similar levels to those reported by Crespo et al. [16]. The mineral profile was mainly dominated by five major elements—K, P, Ca, Na, and Mg—with potassium showing the highest concentrations (average content of 2834.5 mg kg^{-1}). Phosphorous, calcium magnesium, and sodium were also present in high concentrations, representing approximately 20% of the total mineral content, while other elements (Fe, Cu, Zn, and Sr) accounted for less than 1% of the mineral profile. It should be noted that these results are in line with previous findings [7].

Multivariate analysis of variance (MANOVA) was applied to test the effects of the cultivar and cultivation system on the chemical profile, and analysis of variance (ANOVA) with a Tukey HSD post hoc test was used to evaluate the statistical significance of the differences for each compound or element measured. The multivariate test showed that both factors have a significant effect on the content of sugars, organic acids, and polyphenols ($p < 0.001$), but not on the mineral profile ($p > 0.1$ and $p > 0.5$ for the variety and cultivation system, respectively). Univariate results for each variable are shown in Table 1. "Camarosa" and "Ventana" were found to be the richest cultivars in total sugars and organic acids. In particular, "Camarosa" strawberries showed the highest content of sucrose and malic acid. The "Ventana" cultivar presented the richest profile in phenolic acids, mainly dominated by ellagic acid, while "Camarosa" and "Aromas" varieties showed higher concentrations of total polyphenols, mainly anthocyanins.

Table 1. Concentrations (expressed as the mean ± standard deviation) of sugars (g kg^{-1}), organic acids (g kg^{-1}), phenolic compounds (mg kg^{-1}), and mineral elements (mg kg^{-1}) in each strawberry cultivar, and *p* values obtained by ANOVA.

Compounds	Aromas	Camarosa	Diamante	Medina	Ventana	*p* Value
sucrose	6.9 ± 4.8	14.1 ± 2.3	9.2 ± 1.7	6.7 ± 3.4	10.0 ± 3.1	0.0003
glucose	11.9 ± 4.4	11.7 ± 3.2	12.4 ± 3.0	12.3 ± 3.9	14.6 ± 3.9	0.4836
fructose	11.6 ± 4.1	11.0 ± 2.8	11.5 ± 2.4	11.4 ± 3.8	13.3 ± 3.4	0.6800
Total Sugars	**30.5 ± 12.8**	**36.7 ± 6.3**	**33.1 ± 5.7**	**30.4 ± 10.2**	**37.9 ± 8.7**	**0.3383**
ascorbic acid	0.1 ± 0.04	0.2 ± 0.1	0.2 ± 0.02	0.2 ± 0.08	0.2 ± 0.1	0.4145
citric acid	5.1 ± 2.0	6.3 ± 0.8	5.3 ± 1.1	4.7 ± 1.4	5.3 ± 0.8	0.1937
tartaric acid	0.08 ± 0.08	0.1 ± 0.04	0.2 ± 0.06	0.07 ± 0.09	0.2 ± 0.07	0.0959
malic acid	0.5 ± 0.1	2.4 ± 0.4	0.6 ± 0.2	0.5 ± 0.1	0.7 ± 0.2	0.0871
Total Acids	**5.8 ± 2.2**	**8.9 ± 2.9**	**6.2 ± 1.3**	**5.5 ± 1.5**	**6.4 ± 0.8**	**0.0074**
pelargonidin derivative 1	0.9 ± 0.3	0.8 ± 0.2	0.6 ± 0.2	0.7 ± 0.3	0.7 ± 0.4	0.0750
cyanidin 3-glucoside	6.4 ± 1.6	4.0 ± 0.8	3.0 ± 1.1	3.8 ± 0.2	1.4 ± 0.6	0.0000
pelargonidin 3-glucoside	120.9 ± 17.7	117.2 ± 29.9	72.4 ± 3.3	102.7 ± 30.9	86.1 ± 22.2	0.0003
pelargonidin 3-rutinoside	7.4 ± 1.6	15.8 ± 5.1	5.2 ± 1.0	6.2 ± 0.7	6.7 ± 2.5	0.0000
pelargonidin derivative 2	0.7 ± 0.2	0.6 ± 0.3	0.6 ± 0.2	0.7 ± 0.09	0.8 ± 0.4	0.8180
pelargonidin acetate	3.0 ± 0.4	2.3 ± 0.7	1.4 ± 0.2	2.1 ± 0.8	1.1 ± 0.4	0.0000
Total Anthocyanins	**139.4 ± 18.3**	**140.9 ± 35.9**	**83.3 ± 1.9**	**116.2 ± 32.3**	**96.7 ± 25.1**	**0.0001**
p-hydroxybenzoic acid	0.6 ± 0.1	1.4 ± 0.3	0.8 ± 0.9	0.3 ± 0.02	0.5 ± 0.03	0.0139
caffeic acid	0.4 ± 0.1	0.6 ± 0.2	0.2 ± 0.01	0.5 ± 0.1	0.9 ± 0.2	0.0001
p-coumaric acid	7.8 ± 1.7	6.6 ± 3.1	4.2 ± 2.2	5.8 ± 1.3	19.3 ± 6.1	0.0000
ferulic acid	0.08 ± 0.02	0.2 ± 0.07	0.2 ± 0.02	0.1 ± 0.04	0.4 ± 0.08	0.0078
ellagic acid	39.3 ± 10.3	35.8 ± 10.5	54.3 ± 23.8	45.8 ± 22.3	63.4 ± 25.9	0.1295
Total Phenolic Acids	**48.1 ± 11.1**	**44.6 ± 13.5**	**59.8 ± 24.5**	**52.6 ± 22.4**	**84.4 ± 36.3**	**0.0129**
quercetin	1.4 ± 0.08	1.5 ± 0.2	0.9 ± 0.2	0.9 ± 0.3	0.7 ± 0.1	0.0249
Kaempferol O-glucoside	23.0 ± 7.2	29.5 ± 10.3	18.3 ± 6.0	21.2 ± 9.5	30.3 ± 8.3	0.0262
Total Flavonols	**24.4 ± 7.2**	**31.0 ± 10.4**	**19.3 ± 6.1**	**22.2 ± 9.9**	**31.0 ± 7.9**	**0.0237**
P	224.3 ± 35.6	251.6 ± 11.3	196.2 ± 29.5	219.2 ± 21.8	218.5 ± 16.4	0.0025
Ba	0.6 ± 0.2	0.5 ± 0.06	0.4 ± 0.06	0.4 ± 0.02	0.4 ± 0.04	0.9586
Ca	210.7 ± 24.1	244.2 ± 35.4	156.4 ± 20.1	195.4 ± 33.2	235.2 ± 28.6	0.9732
Cr	0.1 ± 0.02	0.06 ± 0.01	0.06 ± 0.02	0.05 ± 0.01	0.2 ± 0.03	0.9932
Cu	4.6 ± 1.9	4.8 ± 1.1	4.8 ± 1.4	4.9 ± 1.4	5.4 ± 1.3	0.9915
Fe	7.8 ± 1.2	7.6 ± 1.1	5.8 ± 1.6	8.7 ± 1.1	7.5 ± 1.9	0.9723
K	2843.5 ± 287.6	2788.2 ± 357.4	2098.1 ± 210.2	2844.8 ± 351.9	3597.9 ± 443.5	0.9263
Mg	226.7 ± 26.9	179.3 ± 26.0	142.5 ± 20.1	167.5 ± 23.7	222.6 ± 30.7	0.9526
Mn	8.8 ± 1.9	6.9 ± 1.0	5.9 ± 1.9	6.8 ± 1.1	9.9 ± 1.1	0.9394
Na	189.1 ± 28.9	116.0 ± 23.7	98.3 ± 21.5	88.9 ± 17.4	126.2 ± 16.1	0.9000
Ni	0.3 ± 0.06	0.3 ± 0.02	0.3 ± 0.07	0.3 ± 0.03	0.3 ± 0.05	0.9890
Sr	6.0 ± 1.0	3.6 ± 1.7	3.1 ± 1.5	4.8 ± 1.8	5.6 ± 1.7	0.8802
Zn	3.2 ± 0.9	7.5 ± 0.8	3.74 ± 0.49	3.47 ± 0.33	4.26 ± 0.30	0.5551

ANOVA, One-way analysis of variance.

3.2. Application of Pattern Recognition Tools for Selecting Chemical Descriptors of Cultivar and Agronomic Conditions

Several chemometric techniques, including unsupervised and supervised pattern recognition procedures, were employed to achieve a reliable differentiation between strawberry samples according to the cultivar, cultivation system, and/or campaign.

A preliminary data exploration was carried out by principal component analysis (PCA), using autoscaled data and only considering the principal components (PCs) with eigenvalues greater than 1. This PCA model allowed 84% of the total variance to be explained with five components. As shown in the scores plot built using the two first principal components (Figure 1A), a clear separation was observed along the PC1 among samples collected in the two consecutive campaigns. The first PC explained 28% of the variance, and was positively related to fructose and tartaric acid, and negatively associated with pelargonidin 3-glucoside, total flavonoids, and total polyphenols. That is, the content of anthocyanins and total polyphenols was greater during the second campaign, when rainfall, and maximum and minimum temperatures were higher, whereas fructose and tartaric acid contents were more abundant in the first campaign, when total radiation was higher. In this vein, it has previously been described that the content of many phenolic compounds and the antioxidant capacity increase in berry fruits as the temperature increases [17]. Moreover, a low light intensity and high temperatures have also been demonstrated to provoke a decreased synthesis of sugars and ascorbic acid [7,11,18].

On the other hand, the plotting of the second and fourth PCs provided a certain differentiation, depending on the cultivar (Figure 1B), with "Camarosa" and "Aromas" varieties distributed on the left side of the projection, and the rest of the samples located on the right side. The most relevant compounds contributing to this separation were anthocyanins (increased in "Camarosa" and "Aromas") and phenolic acids (decreased in "Camarosa" and "Aromas)", in accordance with the results obtained by ANOVA.

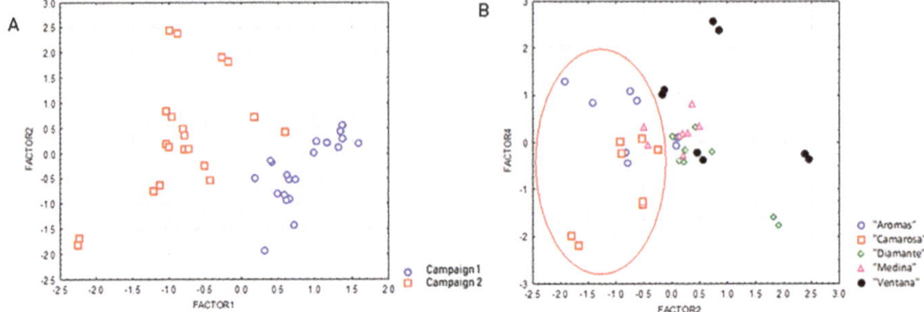

Figure 1. Principal component analysis (PCA) score plots showing the projection of strawberry samples in the plane defined by the following principal components: (**A**) PC1 vs. PC2, separation of samples according to the campaign; (**B**) PC2 vs. PC4, separation of samples according to the cultivar.

After this preliminary data exploration, several supervised chemometric tools were employed to build classification models with the aim of assessing the potential of the multi-chemical profile investigated in this work to authenticate strawberries according to the variety and cultivation conditions. For this purpose, multiple supervised pattern recognition procedures have recently been proposed in food research to solve authentication problems for various foods with a high commercial value, such as strawberry [11,15], olive oil [19–21], or wine [22,23]. In the present study, three complementary statistical techniques were tested: linear discriminant analysis (LDA), soft independent modeling of class analogy (SIMCA), and partial least squares discriminant analysis (PLS-DA).

Linear discriminant analysis (LDA) was first applied to all the study variables, yielding a model capable of explaining 96% of the total variance with a 95% prediction ability. Applying forward stepwise analysis, cyanidin 3-glucoside, pelargonidin 3-rutinoside, p-coumaric acid, phosphorous, malic acid, caffeic acid, and quercetin were identified as the most discriminant variables among cultivars. As shown in Figure 2A, all samples were correctly classified, with the exception of two samples of "Medina" cultivar, which were classified as "Diamante". In line with the results from PCA, "Aromas" and "Camarosa" cultivars were clearly differentiated from the rest of the samples along the first root, while the second one described almost complete separation between the other three cultivars.

Soft independent modeling of class analogy (SIMCA) was subsequently applied to the same data matrix used in LDA, with the aim of looking for possible overlap among the study groups. Using a seven-fold cross-validation procedure, 3-PC-based models were obtained explaining 96.4%, 94.5%, 95.0%, 92.5%, and 97.2% of variance for the classes "Aromas", "Camarosa", "Diamante", "Medina", and "Ventana", respectively. These models also provided very good results in terms of their prediction ability, with 86.7%, 81.5%, 80.8%, 71.3%, and 88.5% correct prediction for the five cultivars. In this line, representation of the corresponding Coomans plot showed a correct classification of strawberries according to the variety based on their chemical composition (Figure 2B). However, SIMCA modeling did not provide suitable results for the classification of strawberry samples according to agronomic conditions, with samples appearing in the overlapping area from the Coomans plots (figure not shown).

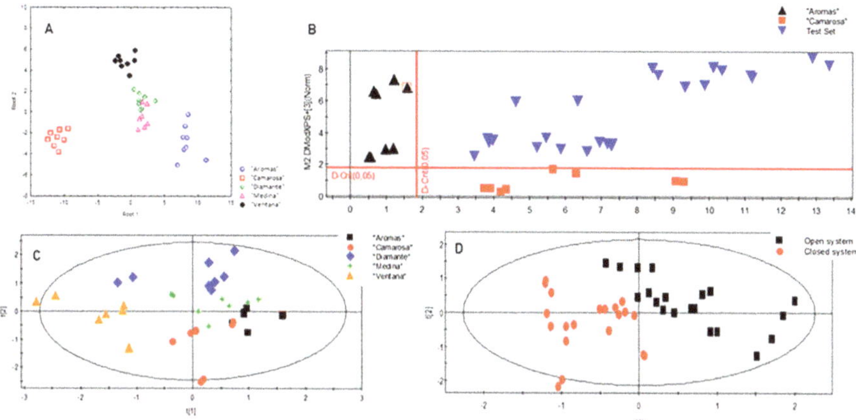

Figure 2. Results obtained from supervised chemometric modeling. (**A**) Linear discriminant analysis (LDA) scores plot showing the distribution of samples in the plane defined by the two first principal components using the cultivar as the categorical variable; (**B**) Soft independent model class analogy (SIMCA) Coomans plots for the classification of strawberry samples according to the cultivar: "Aromas" vs. "Camarosa"; (**C**) Partial least squares discriminant analysis (PLS-DA) scores plot showing the distribution of samples in the plane defined by the two first principal components using the cultivar as the categorical variable; (**D**) PLS-DA scores plot showing the distribution of samples in the plane defined by the two first principal components using the cultivation system as the categorical variable.

Finally, partial least squares discriminant analysis (PLS-DA) was also employed as a more powerful technique for class differentiation and for the selection of the most discriminant variables. A five-component model was obtained with a good quality of fit ($R^2_X = 0.744$) and predictive ability ($Q^2 = 0.413$) for the classification of strawberry samples according to the cultivar (Figure 2C). The most important chemical descriptors driving this separation were anthocyanins and phenolic acids, in line with previous findings from ANOVA and LDA. Interestingly, PLS-DA modeling also enabled the discrimination of samples grown in the two cultivation systems (i.e., open and closed soilless systems). The PLS-DA model explained 70.2% of the variance (Figure 2D), with p-hydroxybenzoic acid, ferulic acid, unknown derivatives of pelargonidin, glucose, pelargonidin acetylglucoside, and cyanidin 3-glucoside being the most discriminant variables.

4. Conclusions

In this work, we have evaluated the potential of combining multi-chemical profiling and complementary statistical techniques to investigate the effect of the genotype and cultivation conditions on the chemical composition of strawberry fruits. The five cultivars investigated showed clear differences in the content of anthocyanins, phenolic acids, sucrose, and malic acid. On the other hand, climatic conditions (e.g., rainfall and temperature) were responsible for slight changes in the polyphenolic profile, with an increased content of anthocyanins and total polyphenols in strawberry fruits grown under higher rainfall and more extreme temperatures. Similarly, the cultivation conditions (i.e., open/closed soilless system) also induced minor changes in concentrations of several anthocyanins and phenolic acids. The present work therefore demonstrates that multi-chemical profiling can be used to differentiate among strawberry cultivars grown under different agronomic conditions, thus showing a great applicability for food traceability. In future studies, this approach could also be tested to search for characteristic patterns associated with the geographical origin, ripeness status, and other factors related to food production.

Author Contributions: Conceptualization, Á.F.-R.; methodology, A.S. and Á.F.-R.; software, R.G.-D. and Á.F.-R.; validation, R.G.-D. and Á.F.-R.; formal analysis, I.A.; investigation, R.G.-D., A.S., I.A., and Á.F.-R.; resources, A.S. and Á.F.-R.; data curation, Á.F.-R.; writing—original draft preparation, Á.F.-R.; writing—review and editing, R.G.-D., A.S., I.A., and Á.F.-R.; visualization, R.G.-D. and Á.F.-R.; supervision, Á.F.-R.; project administration, Á.F.-R. All authors have read and agreed to the published version of the manuscript.

Funding: This research received no external funding.

Conflicts of Interest: The authors declare no conflict of interest.

References

1. Aung, M.M.; Chang, Y.S. Traceability in a food supply chain: Safety and quality perspectives. *Food Control* **2014**, *39*, 172–184. [CrossRef]
2. Giampieri, F.; Forbes-Hernandez, T.Y.; Gasparrini, M.; Alvarez-Suarez, J.M.; Afrin, S.; Bompadre, S.; Quiles, J.L.; Mezzetti, B.; Battino, M. Strawberry as a health promoter: An evidence based review. *Food Funct.* **2015**, *6*, 1386–1398. [CrossRef] [PubMed]
3. Basu, A.; Nguyen, A.; Betts, N.M.; Lyons, T.J. Strawberry as a functional food: An evidence-based review. *Crit. Rev. Food Sci. Nutr.* **2014**, *54*, 790–806. [CrossRef] [PubMed]
4. Bagchi, D.; Nair, S. *Developing New Functional Food and Nutraceutical Products*, 1st ed.; Academic Press: San Diego, CA, USA, 2016.
5. Sturm, K.; Koron, D.; Stampar, F. The composition of fruit of different strawberry varieties depending on maturity stage. *Food Chem.* **2003**, *83*, 417–422. [CrossRef]
6. Giampieri, F.; Tulipani, S.; Alvarez-Suarez, J.M.; Quiles, J.L.; Mezzetti, B.; Battino, M. The strawberry: Composition, nutritional quality, and impact on human health. *Nutrition* **2012**, *28*, 9–19. [CrossRef] [PubMed]
7. Akhatou, I.; Fernández-Recamales, A. Influence of cultivar and culture system on nutritional and organoleptic quality of strawberry. *J. Sci. Food Agric.* **2014**, *94*, 866–875. [CrossRef]
8. Nowicka, A.; Kucharska, A.Z.; Sokół-Łętowska, A.; Fecka, I. Comparison of polyphenol content and antioxidant capacity of strawberry fruit from 90 cultivars of *Fragaria* × *ananassa* Duch. *Food Chem.* **2019**, *270*, 32–46. [CrossRef]
9. Giampieri, F.; Alvarez-Suarez, J.M.; Mazzoni, L.; Romandini, S.; Bompadre, S.; Diamanti, J.; Capocasa, F.; Mezzetti, B.; Quiles, J.L.; Ferreiro, M.S.; et al. The potential impact of strawberry on human health. *Nat. Prod. Res.* **2013**, *27*, 448–455. [CrossRef]
10. Akhatou, I.; Fernández-Recamales, A. Nutritional and nutraceutical quality of strawberries in relation to harvest time and crop conditions. *J. Agric Food Chem.* **2014**, *62*, 5749–5760. [CrossRef]
11. Akhatou, I.; González-Domínguez, R.; Fernández-Recamales, Á. Investigation of the effect of genotype and agronomic conditions on metabolomic profiles of selected strawberry cultivars with different sensitivity to environmental stress. *Plant Physiol. Biochem.* **2016**, *101*, 14–22. [CrossRef]
12. Wang, S.Y.; Millner, P. Effect of different cultural systems on antioxidant capacity, phenolic content, and fruit quality of strawberries (*Fragaria* × *ananassa* Duch.). *J. Agric. Food Chem.* **2009**, *57*, 9651–9657. [CrossRef] [PubMed]
13. Kruger, E.; Josuttis, M.; Nestby, R.; Toldam-Andersen, T.B.; Carlen, C.; Mezzetti, B. Influence of growing conditions at different latitudes of Europe on strawberry growth performance, yield and quality. *J. Berry Res.* **2012**, *2*, 143–157. [CrossRef]
14. Aaby, K.; Mazur, S.; Nes, A.; Skrede, G. Phenolic compounds in strawberry (*Fragaria* × *ananassa* Duch.) fruits: Composition in 27 cultivars and changes during ripening. *Food Chem.* **2012**, *132*, 86–97. [CrossRef] [PubMed]
15. Akhatou, I.; Sayago, A.; González-Domínguez, R.; Fernández-Recamales, Á. Application of Targeted Metabolomics to Investigate Optimum Growing Conditions to Enhance Bioactive Content of Strawberry. *J. Agric. Food Chem.* **2017**, *65*, 9559–9567. [CrossRef] [PubMed]
16. Crespo, P.; Giné Bordonaba, J.; Terry, L.A.; Carlen, C. Characterisation of major taste and health-related compounds of four strawberry genotypes grown at different Swiss production sites. *Food Chem.* **2010**, *122*, 16–24. [CrossRef]
17. Wang, S.Y.; Zheng, W. Effect of plant growth temperature on antioxidant capacity in strawberry. *J. Agric. Food Chem.* **2001**, *49*, 4977–4982. [CrossRef]

18. Pereira, G.E.; Gaudillere, J.-P.; Pieri, P.; Hilbert, G.; Maucourt, M.; Deborde, C.; Moing, A.; Rolin, D. Microclimate influence on mineral and metabolic profiles of grape berries. *J. Agric. Food Chem.* **2006**, *54*, 6765–6775. [CrossRef]
19. Sayago, A.; González-Domínguez, R.; Beltrán, R.; Fernández-Recamales, Á. Combination of complementary data mining methods for geographical characterization of extra virgin olive oils based on mineral composition. *Food Chem.* **2018**, *261*, 42–50. [CrossRef]
20. Sayago, A.; González-Domínguez, R.; Urbano, J.; Fernández-Recamales, Á. Combination of vintage and new-fashioned analytical approaches for varietal and geographical traceability of olive oils. *LWT* **2019**, *111*, 99–104. [CrossRef]
21. González-Domínguez, R.; Sayago, A.; Morales, M.T.; Fernández-Recamales, Á. Assessment of Virgin Olive Oil Adulteration by a Rapid Luminescent Method. *Foods* **2019**, *8*, 287. [CrossRef]
22. Versari, A.; Laurie, V.F.; Ricci, A.; Laghi, L.; Parpinello, G.P. Progress in authentication, typification and traceability of grapes and wines by chemometric approaches. *Food Res. Int.* **2014**, *60*, 2–18. [CrossRef]
23. González-Domínguez, R.; Sayago, A.; Fernández-Recamales, Á. Metabolomics: An Emerging Tool for Wine Characterization and the Investigation of Health Benefits. In *Engineering Tools in the Beverage Industry. Volume 3: The Science of Beverages*, 1st ed.; Grumezescu, A., Holban, A.M., Eds.; Woodhead Publishing: Duxford, UK, 2019; Volume 3, pp. 315–350.

MDPI
St. Alban-Anlage 66
4052 Basel
Switzerland
Tel. +41 61 683 77 34
Fax +41 61 302 89 18
www.mdpi.com

Foods Editorial Office
E-mail: foods@mdpi.com
www.mdpi.com/journal/foods